DIFFERENTIAL EQUATIONS

MONOGRAPHS AND TEXTBOOKS IN
PURE AND APPLIED MATHEMATICS

1. *K. Yano,* Integral Formulas in Riemannian Geometry (1970)
2. *S. Kobayashi,* Hyperbolic Manifolds and Holomorphic Mappings (1970)
3. *V. S. Vladimirov,* Equations of Mathematical Physics (A. Jeffrey, ed.; A. Littlewood, trans.) (1970)
4. *B. N. Pshenichnyi,* Necessary Conditions for an Extremum (L. Neustadt, translation ed.; K. Makowski, trans.) (1971)
5. *L. Narici et al.,* Functional Analysis and Valuation Theory (1971)
6. *S. S. Passman,* Infinite Group Rings (1971)
7. *L. Dornhoff,* Group Representation Theory. Part A: Ordinary Representation Theory. Part B: Modular Representation Theory (1971, 1972)
8. *W. Boothby and G. L. Weiss, eds.,* Symmetric Spaces (1972)
9. *Y. Matsushima,* Differentiable Manifolds (E. T. Kobayashi, trans.) (1972)
10. *L. E. Ward, Jr.,* Topology (1972)
11. *A. Babakhanian,* Cohomological Methods in Group Theory (1972)
12. *R. Gilmer,* Multiplicative Ideal Theory (1972)
13. *J. Yeh,* Stochastic Processes and the Wiener Integral (1973)
14. *J. Barros-Neto,* Introduction to the Theory of Distributions (1973)
15. *R. Larsen,* Functional Analysis (1973)
16. *K. Yano and S. Ishihara,* Tangent and Cotangent Bundles (1973)
17. *C. Procesi,* Rings with Polynomial Identities (1973)
18. *R. Hermann,* Geometry, Physics, and Systems (1973)
19. *N. R. Wallach,* Harmonic Analysis on Homogeneous Spaces (1973)
20. *J. Dieudonné,* Introduction to the Theory of Formal Groups (1973)
21. *I. Vaisman,* Cohomology and Differential Forms (1973)
22. *B.-Y. Chen,* Geometry of Submanifolds (1973)
23. *M. Marcus,* Finite Dimensional Multilinear Algebra (in two parts) (1973, 1975)
24. *R. Larsen,* Banach Algebras (1973)
25. *R. O. Kujala and A. L. Vitter, eds.,* Value Distribution Theory: Part A; Part B: Deficit and Bezout Estimates by Wilhelm Stoll (1973)
26. *K. B. Stolarsky,* Algebraic Numbers and Diophantine Approximation (1974)
27. *A. R. Magid,* The Separable Galois Theory of Commutative Rings (1974)
28. *B. R. McDonald,* Finite Rings with Identity (1974)
29. *J. Satake,* Linear Algebra (S. Koh et al., trans.) (1975)
30. *J. S. Golan,* Localization of Noncommutative Rings (1975)
31. *G. Klambauer,* Mathematical Analysis (1975)
32. *M. K. Agoston,* Algebraic Topology (1976)
33. *K. R. Goodearl,* Ring Theory (1976)
34. *L. E. Mansfield,* Linear Algebra with Geometric Applications (1976)
35. *N. J. Pullman,* Matrix Theory and Its Applications (1976)
36. *B. R. McDonald,* Geometric Algebra Over Local Rings (1976)
37. *C. W. Groetsch,* Generalized Inverses of Linear Operators (1977)
38. *J. E. Kuczkowski and J. L. Gersting,* Abstract Algebra (1977)
39. *C. O. Christenson and W. L. Voxman,* Aspects of Topology (1977)
40. *M. Nagata,* Field Theory (1977)
41. *R. L. Long,* Algebraic Number Theory (1977)
42. *W. F. Pfeffer,* Integrals and Measures (1977)
43. *R. L. Wheeden and A. Zygmund,* Measure and Integral (1977)
44. *J. H. Curtiss,* Introduction to Functions of a Complex Variable (1978)
45. *K. Hrbacek and T. Jech,* Introduction to Set Theory (1978)
46. *W. S. Massey,* Homology and Cohomology Theory (1978)
47. *M. Marcus,* Introduction to Modern Algebra (1978)
48. *E. C. Young,* Vector and Tensor Analysis (1978)
49. *S. B. Nadler, Jr.,* Hyperspaces of Sets (1978)
50. *S. K. Segal,* Topics in Group Kings (1978)
51. *A. C. M. van Rooij,* Non-Archimedean Functional Analysis (1978)
52. *L. Corwin and R. Szczarba,* Calculus in Vector Spaces (1979)

Additional Volumes in Preparation

DIFFERENTIAL EQUATIONS

Introduction and Qualitative Theory

Second Edition, Revised and Expanded

Jane Cronin
Rutgers University
New Brunswick, New Jersey

Marcel Dekker, Inc.   New York•Basel•Hong Kong

Library of Congress Cataloging-in-Publication Data

Cronin, Jane
  Differential equations : introduction and qualitative theory /
Jane Cronin. — 2nd ed., rev. and expanded.
    p. cm. — (Monographs and textbooks in pure and applied
mathematics ; 180)
  Includes bibliographical references (p. 357–369) and index.
  ISBN 0-8247-9189-4 (acid–free)
  1. Differential equations. I. Title. II. Series.
QA372.C92 1994
515'.352—dc20   93-46027
                                                                       CIP

The publisher offers discounts on this book when ordered in bulk quantities. For more information, write to Special Sales/Professional Marketing at the address below.

This book is printed on acid-free paper.

MARCEL DEKKER, INC.
270 Madison Avenue, New York, New York 10016

Current printing (last digit):
10 9 8 7 6 5 4 3 2 1

PRINTED IN THE UNITED STATES OF AMERICA

Preface to Second Edition

The main purpose of the changes in this edition is to make the book more effective as a textbook. To this end, the following steps have been taken.

First, the material in Chapter 1 has been rearranged and augmented to make it more readable. Motivations for studying existence theorems and extension theorems are given. The exercises at the end of the chapter have been extended to illustrate the theory more extensively, and the examples from chemistry and biology introduced at the end of the chapter are treated in detail in the solutions manual. Also the exposition and organization of the chapter have been reworked with a view to clarification.

In Chapter 2, a treatment of the Sturm-Liouville theory has been added. The theory has been treated in outline form with specific references for proofs. The purpose of the discussion is to make clear the relation between initial value problems and boundary value problems, to describe how the Sturm-Liouville theory is rooted in elementary linear algebra, and to show that the Sturm-Liouville problem and its solution form a concrete example in functional analysis. (Unlike the other material in this book, the Sturm-Liouville problem is an infinite-dimensional problem and really belongs in a course in functional analysis.)

The text in all the chapters has been reviewed and in some parts clarified and improved. (The proof of the Instability Theorem has been substantially simplified by allowing use of Lyapunov theory.)

Finally, a solutions manual is available for instructors. The solutions manual (of more than 100 pages) includes detailed solutions for almost all the exercises at the ends of the chapters and also a detailed analysis of the examples given at the end of Chapter 1: the Volterra population equations, the Hodgkin-Huxley nerve conduction equations, the Field-Noyes model of the Belousov-Zhabotinsky reaction, and the Goodwin model of cell activity. The analysis given for these examples that uses the theory from Chapter 1 shows the student the importance of the existence and extension theorems and also provides an early introduction to the use of geometric methods. As the theory is developed in later chapters, it is applied to these examples in the exercises at the ends of the chapters. Thus fairly extensive analyses of these examples are given in the solutions manual.

The solutions in the manual are written in detail so that the instructor can easily judge the length of an outside assignment and so that the students can read them. Thus if time limitations prevent discussion in class of a problem (an unfortunately frequent event in my experience) copies of solutions can be distributed to students.

I want to thank the faculty and students whose comments have helped greatly in the preparation of this book.

I am deeply indebted to Dottie Phares who prepared the $\mathcal{AMS}$-LaTeX version of this book with great understanding and skill.

Jane Cronin

Preface to First Edition

This book has two objectives: first, to introduce the reader to some basic theory of ordinary differential equations which is needed regardless of the direction pursued in later studies and, second, to give an account of some qualitative theory of ordinary differential equations which has been developed in the last couple of decades and which may be useful in problems which have arisen in quantitiative studies in chemical kinetics, biochemical systems, physiological problems and other biological problems. The only prerequisites for reading the book are the first semester of a course in advanced calculus and a semester of linear algebra.

Chapters 1, 2, 3, 4 and the first part of chapter 6 are a suitable basis for a first course in ordinary differential equations at the advanced undergraduate or beginning graduate level. The discussion differs from the conventional presentation only in that there is heavier emphasis on stability theory and there is no treatment of eigenvalue problems. The reason for emphasizing stability lies in the growing importance of stability in studies of problems in chemistry and biology. While neutral stability may be significant in physics, the unaccounted-for disturbances in chemical and biological systems are sufficiently important to suggest that in these studies, a stronger stability condition, i.e., some kind of asymptotic stability, is needed. Eigenvalue problems have been treated in admirable detail in many books and we have chosen a more limited course (the search for periodic solutions) and have extended only this to the nonlinear case.

Not all of the material in Chapters 1 through 4 needs to be considered. It is certainly not necessary to study in detail all the existence theorems given in Chapter 1, although the functional analysis approach used in some of the proofs helps to orient a student who has some experience in that subject. On the other hand, the extension theorems should be emphasized because they are of crucial importance in understanding the difficulties of working with solutions of nonlinear equations. Because the linear theory is so important in itself and for later work, the material in Chapter 2 should be given strong emphasis. But the treatment of two-dimensional systems in the latter part of Chapter 3 can be merely indicated or omitted. In Chapter 4, the basic results are important, but whether all the details of proofs, for example the proof of the Instability Theorem, should be included is questionable. The proof of the Poincaré-Bendixson Theorem in Chapter 6 is given in full detail and the use of the Jordan Curve Theorem in the proof is described with some care.

Chapter 5 is an introduction to the Lyapunov Second Method. The reasons for using the method and the basic theorem are discussed in detail, but only indications are given of the important applications that have been made of the method. A more complete discussion may be found in the lucid account given by LaSalle and Lefschetz [1963].

The second part of Chapter 6 treats a theorem of George Sell which is a kind of extension of the Poincaré-Bendixson theorem to the n-dimensional case. The Poincaré-Bendixson theorem itself cannot be generalized to the n-dimensional case, but Sell's theorem says roughly that if an n-dimensional autonomous system has a bounded solution with suitable asymptotic stability, then that solution approaches an asymptotically stable periodic solution. From the viewpoint of pure mathematics, this is much less impressive than the Poincaré-Bendixson theorem because the strong hypothesis is imposed that an asymptotically stable solution exists. From the point of view of applications in chemistry and biology where it seems highly likely that only solutions with some kind of asymptotic stability are significant, Sell's result is very important, for it suggests that the search for periodic solutions should be supplanted by a search for asymptotically stable solutions.

Chapter 7 is a fairly detailed treatment of the classical problem of branching or bifurcation of periodic solutions if a small parameter is varied. There is an enormous literature on various aspects of this problem, and our aim here is to provide a general discussion of the qualitative aspects of the problem, i.e., the existence and stability of periodic solutions is treated. For the nonautonomous case, a general discussion along the lines given by Malkin [1959] is given. The treatment is somewhat more streamlined than Malkin's and is completed by applying topological degree theory to obtain explicit (i.e., computable) sufficient conditions for the existence and stability of periodic solutions. For the autonomous case, the treatment given by Coddington and Levinson [1952] and the Hopf Bifurcation Theorem are discussed and extended by using degree theory.

I would like to express my gratitude to the U.S. Army Research Office for partial support during the writing of this book.

In preliminary form, the book was used as a text in several courses in differential equations. I would like to express my thanks to the many students whose comments improved the text considerably, to Mary Anne Jablonski and Lynn Braun for their efficiency and patient good humor when they typed the various versions of the text, and to Edmund Scanlon who made the drawings for this book.

Jane Cronin

Contents

DIFFERENTIAL EQUATIONS

Chapter 0

Introduction

In freshman and sophomore calculus, the study of differential equatioins is begun by developing a number of more or less computational techniques such as separation of variables. For example, suppose we consider the eqution

$$\frac{dx}{dt} = tx + 5x$$

We "separate variables" and integrate the equation by carrying out the following steps:

$$\frac{dx}{dt} = (t + 5)x$$

$$\frac{dx}{x} = (t + 5)dt$$

$$ln|x| = \frac{t^2}{2} + 5t + C$$

$$|x| = K \exp\left(\frac{t^2}{2} + 5t\right)$$

where C is an arbitrary constant of integration and $K = e^C$ so that K is an arbitrary positive constant. These formal steps are more or less familiar to the student who has taken sophomore calculus or an elementary course in differential equations. A rigorous justification for these steps is less familiar (Exercise 1 in Chapter 1).

It is natural to think that a more advanced study of differential equations consists in the extension and refinement of such techniques as separation of variables. Actually, however, we must take a different direction. The reason is this: although methods like separation of variables are effective when they can be applied, they are applicable only to a very limited class of differential equations, and despite the strenuous efforts of many mathematicians over a long period of time (the eighteenth and nineteenth centuries), there is no indication that these methods can be extended beyond a very limited class of differential equations. The direction that we follow instead is to establish existence theorems for differential equations, i.e., to prove that solutions of a differential equation exist even though we may not be able to compute them explicitly. (The proofs of the existence theorems will suggest methods for finding approximations to solutions but will not, in general, show us how to write the solutions explicitly.) The existence theorems apply to wide classes of differential equations. Thus we give up hope of obtaining explicitly written solutions but gain the advantage of being able to study large classes of equations.

Having obtained the existence of solutions, we proceed then to study some of the properties of the solutions. First we consider such basic characteristics as differentiability of the solution and its continuity with respect to a parameter in the differential equation. Next we study the theory of linear systems which is essential for later work. After these basic studies have been made, there are several courses which can be followed. We choose to study some so-called qualitative properties of solutions: stability, periodicity and almost periodicity. There are several reasons for pursuing this direction. The most important reason is that such study yields a coherent, aesthetically pleasing theory which has important applications in the physical and life sciences.

Qualitative theory of solutions of differential equations originates in the giant developments due to Poincaré and Lyapunov. One of the marks of true mathematical genius is the ability to ask the right questions. The greatly gifted mathematician formulates "good questions," i.e., the questions whose study leads to the development of mathematical theory which is beautiful, profound and useful to people in other fields. In their work on differential equations, Poincaré and Lyapunov displayed this to an extraordinary degree. Both perceived that much of the future development of differential equations would be in the direction of qualitative studies. Indeed, both initiated (independently and from different viewpoints) qualitative studies. Today, when topological notions are familiar to beginning students of mathematics, the idea of a qualitative study may not seem striking. At the time when Poincaré and Lyapunov were working, the introduction of such ideas required tremendous intellectual force and originality. It is scarcely necessary to point out the extent to which Poincaré's work has influenced twentieth century mathematics. Lyapunov's work has had a lesser influence on the development of pure mathematics, but his work on stability foreshadowed twentieth century developments in applied mathematics, the physical sciences and the life sciences to an uncanny extent. To see why this is true, it is necessary to keep in mind that Lyapunov's and Poincaré's work on differential equations was motivated by problems in celestial mechanics. By the 1920's the efforts to develop a theory for the burgeoning subject of radio circuitry led to intensive study of new classes of nonlinear differential equations, and it was realized that the qualitative theory developed by Poincaré [1881, 1882, 1885, 1886, 1892-99] was applicable to these equations. Lyapunov's stability theory (Lyapunov [1892], LaSalle and Lefschetz [1961]), however, remained largely disregarded. It was useful but did not play a crucial role in the study of radio circuit problems (see Andronov and Chaikin [1949]), and its use in celestial mechanics was limited. This limitation stemmed partly from the fact that the stability problems in celestial mechanics were extremely difficult. For example, to prove that certain solutions of the three-body problem are stable in the sense of Lyapunov was only proved in 1961 by Leontovich [1962] who used profound results due to Kolmogorov and Arnold. Indeed, Wintner [1947, p. 98] dismissed Lyapunov's definition of stability as unrealistic because it was too strong a condition. Another reason for the limited use of Lyapunov stability theory in celestial mechanics is the fact that the concept of asymptotic stability introduced by Lyapunov is not applicable in Hamiltonian systems. (The systems of differential equations which occur in celestial mechanics are special cases of Hamiltonian systems.) Thus the Lyapunov theory lay dormant, attended only by some Russian mathematicians, until the advent of control theory in the years following World War II. Efforts to develop a mathematical control theory led to the realization that the Lyapunov theory was well-designed for such studies and there was a widespread growth of interest in stability theory. More recently, it has become clear that if a biological problem can be formulated in terms of a system of ordinary differential equations, then stability theory must play an important role in the study of the system. The reason for this is that since biological systems tend to be quite complicated, the differential equation which is used to describe the system is only a rather crude approximate description. It must be assumed that disturbances of the system (as described by the differential equation) are constantly occurring. This suggests that only those solutions of the differential equation which have strong stability properties are biologically significant, i.e., describe phenomena which actually occur in the biological system.

In choosing to study the qualitative properties described above, we are disregarding very important theory which is useful in the study of physical and biological problems. The crucially important subject of numerical or computational methods for solving differential equations will be omitted. (As will be pointed out later, our choice of a proof for the basic existence theorem is influenced by the wish to indicate the underlying idea of one numerical procedure.)

Chapter 1

Existence Theorems

What This Chapter Is About

We shall be concerned with the existence and uniqueness of solutions and the size of domains of solutions. (We use the word "solution" here to refer to a solution of the initial value problem. See page 12 for a description of the initial value problem.) Since we shall obtain no practical procedures for calculating solutions, it is natural for the student to feel dubious about the value of this material. Would it not be better to omit this theoretical stuff and get on with the real problem of finding explicit solutions? Besides, those who have solved differential equations by using a computer know full well that this is accomplished without any talk about existence, uniqueness, etc.

These are reasonable, serious questions deserving of careful answers. First of all, using a computer to solve a differential equation means that a particular program, code or software is used. The design of the software is based on numerical analysis of differential equations, and this numerical analysis is, in turn, based on the theory to be described here. So if one uses a computer program, this means that somebody else has thrashed out the theory beforehand.

Second, the topics in this chapter are essential to both the structure of the theory of differential equations and the applications of differential equations in physics, engineering, etc. As we proceed in later chapters, we shall see the importance of this material to theory.

Here we shall merely point out why these topics are important for applications.

Certainly it is logical to establish the existence of a solution before attempting to find an explicit formula or approximation for the solution, especially since the question of existence turns out to be somewhat more complicated than might at first be expected. (See Exercises 6, 10 and 11 for untoward outcomes which can occur.) From the point of view of applications, the condition of uniqueness is essential. A solution of a differential equation which models a physical system is used to predict the behavior of the system. If there is more than one solution which satisfies the given initial condition, then prediction is not possible because one does not know which solution to use. Thus if a differential equation is proposed as a model for a physical system and if it can be shown that the solutions of the differential solution are not unique, then the differential equation has questionable value as a model of the physical system. Hence a prudent step before initiating any numerical analysis is to verify that the solutions of the differential equation exist and are unique.

Equally important is the size of the domain of the solution. Very often, the solutions of differential equations that are obtained in a first course in differential equations are defined for all real values of the independent variable. (This frequently occurs because the equation studied is linear.) But it would be wrong to conclude that this is generally the case. As soon as one ventures outside the realm of linear equations, the domain of a solution, i.e., the set of values of the independent variable for which the solution is defined, becomes a serious question. Even for a very simple nonlinear equation, the domain of the solution may be unexpectedly small. As will be shown in Exercise 11 at the end of this chapter, it may be very important to investigate the domain of the solution before undertaking a numerical analysis.

Finally, we want to point out the relative importance of various parts of Chapter 1. The basic existence and uniqueness theorem is Existence Theorem 1.1 (Picard Theorem) of this chapter. Anyone who expects to work with differential equations needs to be well-acquainted with this theorem, its results and its limitations. The

Differentiability Theorem and the Existence Theorem for an Equation with a Parameter describe properties of solutions which are also very important for later work.

Existence Theorems 1.2 and 1.4 are somewhat more sophisticated. They are interesting for a reader with some background in functional analysis, but unnecessary for later parts of this book. Existence Theorem 1.3 has more general interest. A weaker hypothesis (continuity) is used to prove existence of solutions (not necessarily unique). The method of proof is the Euler-Cauchy method which is the basis for numerical studies of differential equations. Also the theorem yields a quick approach to the problem of estimating how large the domain of the solution is.

The second major topic in Chapter 1 concerns the size of the domain of the solution (Extension Theorems 1.1, 1.2, 1.3). This is a crucially important topic and requires careful attention both for later theory and for applications. Study of the domain of the solution is especially important because the student's previous experience with differential equations may be misleading. Judging on the basis of the first course in differential equations, the student may be inclined, quite reasonably, to conclude that if the functions which appear in the differential equation are defined, and have, say, continuous second derivatives for all real values of the variables, then the same will be true of the solutions. This is far from true, and getting information about the domains of solutions is often not easy as we shall see.

Existence Theorem By Successive Approximations

First we introduce some notation and terminology. Throughout, all quantities will be real unless otherwise specified. Whenever we speak of Euclidean n-space (including the line or the plane) we will mean real Euclidean n-space. The domains and ranges of all functions will be subsets of real Euclidean spaces.

It is not purposeful to give a formal definition of differential equation and we shall merely say that a *differential equation* or *system of differential equations* is a statement of equality relating a set of functions and their derivatives. We will consider differential equations in which there is just one independent variable relative to which the derivatives are taken. The differential equation is then said to be an *ordinary* differential equation. We study ordinary differential equations of the form

$$x_1' = f_1(t, x_1, \ldots, x_n)$$

$$x_2' = f_2(t, x_1, \ldots, x_n)$$

$$\cdots \tag{i}$$

$$x_n' = f_n(t, x_1, \ldots, x_n)$$

where $x_1, \ldots, x_n$ are real-valued functions of t; $x_1', \ldots, x_n'$ are the derivatives of $x_1, \ldots, x_n$; and $f_1, \ldots, f_n$ are real-valued functions of $t, x_1, \ldots, x_n$. Notice that by considering only equations of the form (i) we are implicitly using the assumption that our equations can be solved for the derivative. Thus we are excluding from our study equations of the form, for example,

$$a(t)x' + F(t, x) = 0$$

where $F(t, x)$ is a real-valued function and the function $a(t)$ is zero for some of the values of t being considered (i.e., the equation cannot be solved for x').

It is useful, indeed almost essential, to write equation (i) in vector form. For this let x denote the n-vector

$$\begin{bmatrix} x_1 \\ \vdots \\ x_n \end{bmatrix}$$

let x' denote the n-vector

$$\begin{bmatrix} x_1' \\ \vdots \\ x_n' \end{bmatrix}$$

and let $f(t, x)$ denote the n-vector

$$
\begin{bmatrix}
f_1(t, x_1 \ldots, x_n) \\
\vdots \\
f_n(t, x_1, \ldots, x_n)
\end{bmatrix}
$$

For our convenience, we recall the following definitions. A vector function $f(t, x)$ is *continuous* on a set A if each of its components is continuous on A. A vector function

$$
x(t) =
\begin{bmatrix}
x_1(t) \\
\vdots \\
x_n(t)
\end{bmatrix}
$$

is *differentiable* on (a, b) if each of these components $x_1(t), \ldots, x_n(t)$ is differentiable on (a, b). The *derivative of the vector function $x(t)$* is

$$
\frac{dx}{dt} =
\begin{bmatrix}
\frac{dx_1}{dt} \\
\vdots \\
\frac{dx_n}{dt}
\end{bmatrix}
$$

A vector function $x(t)$ is *integrable* on $[a, b]$ if each of the components is integrable on $[a, b]$. The *integral* of $x(t)$ on $[a, b]$ is

$$
\int_a^b x(t)dt =
\begin{bmatrix}
\int_a^b x_1(t)dt \\
\vdots \\
\int_a^b x_n(t)dt
\end{bmatrix}
$$

The *norm* of the vector

$$
x =
\begin{bmatrix}
x_1 \\
\vdots \\
x_n
\end{bmatrix}
$$

denoted by $|x|$, is $\sum_{i=1}^n |x|$. (The norm $|x|$ is the most convenient for our use, but there are other norms in Euclidean n-space. The most familiar is $\|x\| = \{\sum_{i=1}^n |x_i|^2\}^{1/2}$. But it is easy to see that for all vectors x,

$$
\|x\| \leq |x|
$$

and

$$|x| \leq \sqrt{n}\, \|x\|$$

(See Exercise 2.) If inequalities of this kind hold, the norms gives rise to the same topology and they are said to be *equivalent*.)

Now (i) can be written in vector notation as

$$x' = f(t, x) \tag{ii}$$

and will be referred to as a differential equation or as a system of differential equations. (The choice of terminology will depend on whether we are primarily concerned with the components of x or x itself.) If x is an n-vector, then (ii) is sometimes called an *n-dimensional system*. (Unless otherwise stated, n will always denote the dimension of the system.) A differential equation such as (ii) in which only first derivatives appear is called a *first-order equation*. If a derivative of nth order of a component of x occurs in the differential equation but no derivative of order higher than n occurs, the differential equation is called an nth *order differential equation* or n^{th} *order system of differential equations*. However, we need not institute a separate study for n^{th} order equations because n^{th} order systems are easily represented as first order systems. To describe this representation we show how it works for an example. Consider the nth order equation

$$x^{(n)} = g(t, x, x^{(1)}, \ldots, x^{(n-1)}) \tag{iii}$$

where $x^{(j)}$ denotes the jth derivative $(j = 1, \ldots, n)$ of a real-valued function of t. Equation (iii) can be represented as a system of first order equations, i.e., as a system of the form (ii) if we let

$$x_1 = x$$
$$x_2 = x'$$
$$x_3 = x''$$
$$\cdots$$
$$x_{n-1} = x^{(n-2)}$$
$$x_n = x^{(n-1)}$$

In this notation, equation (iii) becomes the first order system

$$x'_1 = x_2$$
$$x'_2 = x_3$$
$$\ldots$$
$$x'_{n-1} = x_n$$
$$x'_n = g(t, x_1, x_2, \ldots, x_n)$$

We shall be concerned with the existence of solutions of differential equation and our first step is to make precise the concept of solution.

Definition. 1.1 *The* projection of a point $(t, x_1, \ldots, x_n)$ *on the t-axis is t. The* projection of a set *is the collection of the projections of the points in the set.*

Definition. 1.2 *Let f be an n-vector function defined and continuous on a connected open set D in (t, x)-space (i.e., the Euclidean space of points $(t, x_1, \ldots, x_n)$ where $t, x_1, \ldots, x_n$ are real numbers) and let $I = (a, b)$ be an open interval on the t-axis (I may be finite or infinite, i.e., a may be a number or $-\infty$ and b may be a number or $+\infty$) such that the projection of D on the t-axis contains I. Let $x(t)$ be an n-vector function defined and differentiable on (a, b) such that $(t, x(t)) \in D$ for all $t \in I$. Then $x(t)$ is a solution on I of the differential equation*

$$x' = f(t, x)$$

if for each $t \in I$

$$\frac{d}{dt} x(t) = f[t, x(t)]$$

(Sometimes it will be convenient to refer to solutions defined on an interval J which includes one or both of its endpoints. This will mean that either the solution is defined on an open interval which contains J or that one-sided derivatives are considered at the included endpoints.)

Notice that we do not require that the domain of the solution $x(t)$ be in any sense maximal. That is, there may exist an interval (c, d) which contains (a, b) properly and a solution $y(t)$ on (c, d) of the differential equation such that $y|(a,b) = x$ where $y|(a,b)$ denotes the function $y(t)$ on the domain (a,b). The solution $y(t)$ is then called an *extension* of the solution $x(t)$ and we say that $x(t)$ is *extended* (to the solution $y(t)$). Later we will study the question of whether such extensions exist.

Before introducing our first existence theorem, we make more precise the question of whether there exists a solution. The reason for doing this is that if there is a solution of the differential equation, there is frequently an infinite set of solutions. For example, the differential equation

$$x' = t$$

has the set of solutions

$$x(t) = \frac{t^2}{2} + c$$

where c is an arbitrary constant. Consequently, instead of seeking just any solution of the differential equation, we search for a solution which satisfies a special condition. That is, instead of seeking just any solution, we look for a solution which has a given value x^0 when t has a given value t_0. Such a solution is denoted by $x(t, t_0, x^0)$. We say that the solution $x(t, t_0, x^0)$ *satisfies the initial condition* that it have the value x^0 when $t = t_0$. Writing the initial condition in the form of an equation, we have:

$$x(t_0, t_0, x^0) = x^0$$

Our existence theorems are designed to answer the question: does there exist a solution of the differential equation which satisfies a given initial condition? (This question is sometimes called the *initial value problem*.)

For the basic existence theorem, we use the concept of a Lipschitz condition on a function.

Definition. 1.3 *Suppose function* $f(t,x)$ *has domain* D *in* (t,x)-*space and suppose there exists a constant* $k > 0$ *such that if* (t,x^1), $(t,x^2) \in D$, *then*

$$|f(t,x^1) - f(t,x^2)| \leq k|x^1 - x^2|$$

Then f satisfies a Lipschitz condition *with respect to x in D, and k is a* Lipschitz constant *for f.*

It is clear that if f satisfies a Lipschitz condition with respect to x in D, then for each fixed t, $f(t,x)$ is a continuous function of x. However, for a fixed x, $f(t,x)$ need not be a continuous function of t. For example, let

$$f(t,x) = 1 \qquad \text{for } t \geq 0,\ x \text{ real}$$
$$f(t,x) = 0 \qquad \text{for } t < 0,\ x \text{ real}$$

This function satisfies a Lipschitz condition with respect to x (let k be any positive constant) but is not continuous as a function of t.

A simple sufficient condition that f satisfy a Lipschitz condition is obtained as follows. Suppose that D is an open set in the (t,x)-plane (here x denotes a 1-vector) such that if (t,x^1) and (t,x^2) are in D, then $(t, x^1 + \theta(x^2 - x^1))$, where $0 \leq \theta \leq 1$, is in D and suppose that $\partial f/\partial x$ exists and is bounded on D. Then by the Mean Value Theorem

$$f(t,x^2) - f(t,x^1) = \left\{ \frac{\partial f}{\partial x}[t, x^1 + \theta_1(x^2 - x^1)] \right\} (x^2 - x^1)$$

where $\theta_1 \in (0,1)$. Thus if $|\partial f/\partial x| \leq M$ on D, then f satisfies a Lipschitz condition with respect to x on D with a Lipschitz constant equal to M. Similar considerations can be made if x is an n-vector ($n > 1$) by using the Mean Value Theorem for functions of several variables.

But f may satisfy a Lipschitz condition even if $\partial f/\partial x$ is not defined as the following example shows: let

$$f(t,x) = |x|$$

for all (t,x) in the (t,x)-plane (here x denotes a 1-vector).

Finally a function may be uniformly continuous on D, an open set, and yet not satisfy a Lipschitz condition as shown in the following example. Let

$$f(t,x) = |x|^\alpha$$

where α is a constant such that $\alpha \in (0,1)$, for all (t,x) in the (t,x)-plane. First $f(t,x)$ is continuous and hence uniformly continuous on

$$\{(t,x)|x \in [-1,1], \ t \text{ real}\}$$

but if $x > 0$,

$$|f(t,x) - f(t,0)| = x^\alpha = x(x^{\alpha-1}) = \frac{1}{x^{1-\alpha}}|x - 0|$$

As x approaches zero, $\frac{1}{x^{1-\alpha}}$ increases without bound. Thus $f(t,x)$ does not satisfy a Lipschitz condition.

Now we are ready to prove the basic existence theorem.

Existence Theorem 1.1 (Picard Theorem) *Let D be an open set in (t,x)-space. Let $(t_0, x^0) \in D$ and let a, b be positive constants such that the set*

$$R = \{(t,x) \mid |t - t_0| \le a, \ |x - x^0| \le b\}$$

is contained in D. Suppose function f is defined and continuous on D and satisfies a Lipschitz condition with respect to x in R. Let

$$
\begin{aligned}
M &= \max_{(t,x)\in R} |f(t,x)| \\
A &= \min\left[a, \frac{b}{M}\right]
\end{aligned}
$$

Then the differential equation

$$x' = f(t,x) \tag{1}$$

has a unique solution $x(t, t_0, x^0)$ on $(t_0 - A, t_0 + A)$ such that $x(t_0, t_0, x^0) = x^0$. This solution $x(t, t_0, x^0)$ is such that

$$|x(t, t_0, x^0) - x^0| \le MA$$

for all $t \in (t_0 - A, t_0 + A)$.

Before proceeding to the proof of Existence Theorem 1.1, we make a few remarks concerning the hypotheses of the theorem and indicating which questions it answers and which questions it leaves unresolved. First, as we will see later, continuity of f is sufficient to insure the existence of a solution. The additional condition that f satisfy a Lipschitz condition is really needed only to prove the uniqueness of solution. In the proof of Existence Theorem 1.1 we will use the Lipschitz condition on f to prove existence because the Lipschitz condition makes it possible to use the method of successive approximations. We obtain in this way an elementary, fairly short and rather elegant proof. But in Existence Theorem 1.3, we will obtain existence of a solution by using only the continuity of f.

From the point of view of pure mathematics, it is vitally important to sort out exactly which hypotheses are needed for existence and which for uniqueness. This question is however somewhat less important than might first be thought. The reason is that much of the theory of ordinary differential equations is developed for systems of equations which satisfy hypotheses strong enough so that solutions are unique. In Chapter 2, we will see how useful the uniqueness is for obtaining information about solutions of linear systems. The theory developed for autonomous systems in Chapter 3 rests largely on the uniqueness condition. From the point of view of applications, uniqueness is all important because without uniqueness, the system of ordinary differential equations and its solutions cannot be used to make quantitative predictions about the behavior of a physical system. (Of course, the presence of chaos can be investigated.) Thus for many purposes, it is reasonable to impose at the beginning hypotheses strong enough so as to assure uniqueness of solution.

Existence Theorem 1.1 leaves unanswered two important kinds of questions. First we often need to know whether a solution satisfies additional conditions: for example, is the solution continuous or differentiable with respect to the initial value x^0. Secondly, it is often important to be able to estimate how large a domain the solution has. Existence Theorem 1.1 merely says that the interval $(t^0 - A, t_0 + A)$ is contained in the domain of the solution $x(t, t_0, x^0)$. The solution $x(t, t_0, x^0)$ might have a much larger domain, and we need to obtain criteria for determining if, for example, the domain includes all $t > \bar{t}$, some fixed value. These questions will be dealt with later in Chapter 1.

Now we proceed to the proof of Existence Theorem 1.1.

Lemma 1.1 *A necessary and sufficient condition that the function* $x(t, t_0, x^0)$, *continuous in t, which satisfies the condition*

$$x(t_0, t_0, x^0) = x^0$$

is a solution of (1) on the interval $(t_0 - r, t_0 + r)$ *where* $r > 0$ *is that* $x(t, t_0, x^0)$ *satisfies the equation:*

$$x(t, t_0, x^0) = x^0 + \int_{t_0}^{t} f[s, x(s, t_0, x^0)]ds \tag{2}$$

for $t \in (t_0 - r, t_0 + r)$.

Proof. If $x(t, t_0, x^0)$ satisfies (2), then since f and x are continuous, we can differentiate (2). The result is equation (1). If $x(t, t_0, x^0)$ satisfies (1), i.e., if

$$\frac{dx}{dt} = f[t, x(t, t_0, x^0)]$$

then taking the definite integral from t_0 to t where $t \in (t_0 - r, t_0 + r)$ on both sides of equation, we obtain:

$$\int_{t_0}^{t} \frac{dx}{ds} ds = x(t, t_0, x^0) - x(t_0, t_0, x^0)$$

$$= x(t, t_0, x^0) - x^0 = \int_{t_0}^{t} f[s, x(s, t_0, x^0)]ds \qquad \square$$

The remainder of the proof of the theorem consists of showing that the sequence

$$x_0(t) = x^0$$

$$x_1(t) = x^0 + \int_{t_0}^{t} f[s, x_0(s)]ds$$

$$x_{m+1}(t) = x^0 + \int_{t_0}^{t} f[s, x_m(s)]ds \quad (m = 1, 2, \dots)$$

converges on $[t_0 - A, t_0 + A]$ to a function which is a solution on $(t_0 - A, t_0 + A)$ of (2) and then showing that this solution is unique. We will actually show that the sequence converges on $[t_0, t_0 + A]$ and that the limit function is a solution on $[t_0, t_0 + A]$ of (2). Similar treatment can be made on $[t_0 - A, t_0]$.

Lemma 1.2 *For each m, the function $x_m(t)$ is defined and continuous on $[t_0, t_0 + A]$ and if $t \in [t_0, t_0 + A]$, then*

$$|x_m(t) - x^0| \le M|t - t_0|$$

Proof. If $m = 0$, the statement is obviously true. If the statement is true for $m = q$, then for $t \in [t_0, t_0 + A]$,

$$|x_q(t) - x^0| \le MA \le b$$

Therefore $f[t, x_q(t)]$ is defined for $t \in [t_0, t_0 + A]$. Since $f[t, x_q(t)]$ is a continuous function of t, then

$$x_{q+1}(t) = x^0 + \int_{t_0}^{t} f[s, x_q(s)]ds$$

is defined and continuous. Also

$$|x_{q+1}(t) - x^0| = \left| \int_{t_0}^{t} f[s, x_q(s)]ds \right| \le M(t - t_0) \qquad \square$$

Lemma 1.3 *The sequence $\{x_m(t)\}$ converges uniformly on $[t_0, t_0 + A]$ to a continuous function $x(t)$.*

Proof. We will prove that the series

$$x_0(t) + \sum_{n=0}^{\infty} [x_{n+1}(t) - x_n(t)] \qquad (3)$$

converges uniformly on $[t_0, t_0 + A]$. For $t \in [t_0, t_0 + A]$, let

$$d_n(t) = |x_{n+1}(t) - x_n(t)|$$

Then for each n,

$$d_n(t) = \left| \int_{t_0}^t \{f[s, x_n(s)] - f[s, x_{n-1}(s)]\} ds \right|$$

$$\leq \int_{t_0}^t |f[s, x_n(s)] - f[s, x_{n-1}(s)]| ds$$

$$\leq k \int_{t_0}^t |x_n(s) - x_{n-1}(s)| ds$$

$$= k \int_{t_0}^t d_{n-1}(s) ds$$

where k is a Lipschitz constant for f on R.

Next we obtain an estimate for $d_n(t)$ by induction. By Lemma 1.2, if $t \in [t_0, t_0 + A]$,

$$d_0(t) = |x_1(t) - x_0(t)| \leq M|t - t_0|$$

Assume that if $t \in [t_0, t_0 + A]$

$$d_n(t) \leq \frac{M}{k} \frac{k^{n+1}(t - t_0)^{n+1}}{(n+1)!}$$

Then

$$d_{n+1}(t) \leq k \int_{t_0}^t d_n(s) ds \leq k \frac{M}{k} \frac{k^{n+1}}{(n+1)!} \int_{t_0}^t (s - t_0)^{n+1} ds$$

$$= \frac{M}{k} \frac{k^{n+2}}{(n+1)!} \frac{1}{n+2} (t - t_0)^{n+2}$$

Thus if $t \in [t_0, t_0 + A]$

$$\sum_{n=0}^{\infty} d_n(t) \leq \frac{M}{k} \sum_{n=0}^{\infty} \frac{k^{n+1}(t - t_0)^{n+1}}{(n+1)!}$$

$$\leq \frac{M}{k} \sum_{n=0}^{\infty} \frac{k^{n+1} A^{n+1}}{(n+1)!}$$

$$= \frac{M}{k} \{\exp[kA] - 1\}$$

Thus the uniform convergence of (3) follows from the Weierstrass M-test or by a simple direct argument.   □

The proof of Lemma 1.3 uses the convergence of the exponential series. An even simpler proof which uses the convergence of the geometric series runs as follows. First impose, if necessary, the additional condition that A is small enough so that

$$kA = r < 1$$

Then we have

$$\max_{t \in [t_0, t_0 + A]} |x_{n+1}(t) - x_n(t)|$$

$$\leq \max \int_{t_0}^{t} |f(s, x_n(s)) - f(s, x_{n-1}(s))| ds$$

$$\leq \max \int_{t_0}^{t} k |x_n(s) - x_{n-1}(s)| ds$$

$$\leq kA \max_{s \in [t_0, t_0 + A]} |x_n(s) - x_{n-1}(s)|$$

$$= r \max |x_n(s) - x_{n-1}(s)|$$

and hence

$$\max |x_{n+1}(t) - x_n(t)| \leq r^n \max |x_1(s) - x^0|$$

The uniform convergence follows at once.

As we will see later, there is no significant advantage in using the exponential series in the proof of Lemma 1.3. The advantage of not requiring that kA be less than one is illusory. It turns out that it is not particularly important to try to maximize the interval of existence of the solution when proving the basic existence theorems. The question which requires detailed study is how far the solution can be extended. We consider this question in the Extension Theorems later in this chapter.

Lemma 1.4 *The function $x(t)$ is a solution of (2) such that $x(t_0) = x^0$.*

Proof. First we show that for $t \in [t_0, t_0 + A]$

$$|x(t) - x^0| \leq b$$

and hence that for all $t \in [t_0, t_0 + A]$, $f[t, x(t)]$ is defined. If $t \in [t_0, t_0 + A]$ and if $\varepsilon > 0$, then if m is sufficiently large,

$$|x(t) - x^0| \le |x(t) - x_m(t)| + |x_m(t) - x^0| < \varepsilon + M(t - t_0)$$

Therefore

$$|x(t) - x^0| \le M(t - t_0) \le MA \le b$$

By the Lipschitz condition on f, we have: for $\varepsilon > 0$,

$$\left| \int_{t_0}^{t} \{f[s, x(s)] - f[s, x_m(s)]\} ds \right|$$

$$\le \int_{t_0}^{t} |f[s, x(s)] - f[s, x_m(s)]| ds$$

$$\le k \int_{t_0}^{t} |x(s) - x_m(s)| ds$$

$$\le k\varepsilon(t - t_0) \text{ if } m \text{ is sufficiently large}$$

$$\le k\varepsilon A$$

Therefore

$$\lim_{m \to \infty} \int_{t_0}^{t} f[s, x_m(s)] ds = \int_{t_0}^{t} f[s, x(s)] ds$$

Taking the limit in m on both sides of the equation

$$x_{m+1}(t) = x^0 + \int_{t_0}^{t} f[s, x_m(s)] ds$$

we obtain:

$$x(t) = x^0 + \int_{t_0}^{t} f[s, x(s)] ds \qquad \square$$

Lemma 1.5 *The solution $x(t)$ of (2), which satisfies the initial condition*

$$x(t_0) = x^0$$

is the only solution of (1) which satisfies this initial condition.

Proof. Suppose there exist solutions $x(t)$ and $\bar{x}(t)$ of (2) on an interval $(t_0 - r, t_0 + r)$, where r is a positive number, such that

$x(t_0) = \bar{x}(t_0) = x_0$. We obtain an estimate on $|x(t) - \bar{x}(t)|$ for $t \in [t_0, t_0 + r - \delta]$, where $0 < \delta < r$ and δ is fixed, as follows. Since $x(t), \bar{x}(t)$ are continuous on $[t_0, t_0 + r - \delta]$ for fixed δ there exists $B > 0$ such that, if $t \in [t_0, t_0 + r - \delta]$, then

$$|x(t) - \bar{x}(t)| \le B$$

But

$$|x(t) - \bar{x}(t)| \le \int_{t_0}^t |f[s, x(s)] - f[s, \bar{x}(s)]| ds$$

$$\le k \int_{t_0}^t |x(s) - \bar{x}(s)| ds \tag{4}$$

Therefore

$$|x(t) - \bar{x}(t)| \le kB(t - t_0)$$

Assume that $|x(t) - \bar{x}(t)| \le [k^m/m!] B(t - t_0)^m$, for m a positive integer. Then by (4), $|x(t) - \bar{x}(t)| \le [k^{m+1}/(m+1)!] [B(t - t_0)^{m+1}]$, which is the $(m+2)$th term in the (convergent) series for $Be^{k(t-t_0)}$. Therefore $|x(t) - \bar{x}(t)| < \varepsilon$; hence $x(t) = \bar{x}(t)$ for $t \in [t_0, t_0 + r - \delta]$. Since δ is arbitrarily small, $x(t) = \bar{x}(t)$ for $t \in (t_0, t_0 + r)$. A similar argument holds for $t \in (t_0 - r, t_0)$.   □

We emphasize that the uniqueness result given by Lemma 1.5, although easy to prove, is crucially important both in later development of the theory and in applications of ordinary differential equations.

This completes the proof of Existence Theorem 1.1. (For sketches of typical solutions, see Figure 1.)   □

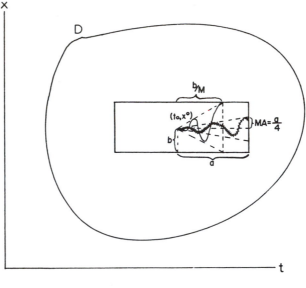

Figure 1

In Figure 1, if $\frac{b}{M} < a$, the solution is defined on

$$[t_0 - \frac{b}{M}, t_0 + \frac{b}{M}] \subseteq [t_0 - a, t_0 + a]$$

and $A = \frac{b}{M}$, and solution $x(t)$ satisfies the condition:

$$|x(t) - x^0| \le M|t - t_0|$$

A possible solution curve if $\frac{b}{M} < a$ is indicated with a solid line in Figure 1. If $\frac{b}{M} \ge a$, the solution is defined on $[t_0 - a, t_0 + a]$. If, for example, $M = \frac{1}{4}$ then $MA = \frac{a}{4}$. A possible solution curve if $\frac{b}{M} \ge a$ is sketched with a "furry" line in Figure 1.

Corollary 1.1 *The solution* $x(t, t_0, x^0)$ *is a continuous function of* (t, x^0)*. That is, given* $\varepsilon > 0$*, then there exists* $\delta > 0$ *such that if* $(t_1, x_1^0), (t_2, x_2^0)$ *are points in the interior of the set* R *and if*

$$|(t_1, x_1^0) - (t_2, x_2^0)| < \delta$$

then

$$|x(t_1, t_0, x_1^0) - x(t_2, t_0, x_2^0)| < \varepsilon.$$

Proof. See Exercise 8.

Corollary 1.1 is an important result in applying differential equations to describe physical and biological systems because the specification of x^0 is only an approximate description of a physical or biological condition.

Differentiability Theorem

It is important sometimes to use the fact that if f has continuous second derivatives then the solution is a differentiable function of x^0, and we proceed to prove this fact now.

Differentiability Theorem. *Let D be an open set in (t, x)-space. Let $(t_0, x^0) \in D$ and suppose function $f(t, x)$ is defined and has continuous second partial derivatives with respect to all variables at each point of D. Then there exist positive numbers a, b such that if $|\bar{x} - x^0| \leq b$, the differential equation*

$$x' = f(t, x) \tag{5}$$

has a unique solution $x(t, t_0, \bar{x})$ on $(t_0 - a, t_0 + a)$ with

$$x(t_0, t_0, \bar{x}) = \bar{x}$$

and such that at each point of the set

$$S = \{(t, x) / |t - t_0| < a, |x - x^0| < b\}$$

the function $x(t, t_0, \bar{x})$ has a continuous third partial derivative with respect to t and has continuous first partial derivatives with respect to each component of $\bar{x}$.

Proof. By Existence Theorem 1.1, there is a solution $x(t, t_0, \bar{x})$ of (5), that is,

$$\frac{\partial x}{\partial t}(t, t_0, \bar{x}) = f[t, x(t, t_0, \bar{x})] \tag{6}$$

Since f is continuous, then $\partial x/\partial t$ is a continuous function of $(t, \bar{x})$. Since f has continuous first and second partial derivatives, then using the chain rule to differentiate (6), we find that $\partial^2 x/\partial t^2$ and $\partial^3 x/\partial t^3$ are continuous functions of $(t, \bar{x})$.

By Lemma 1.1, we know that if $x(t, t_0, \bar{x})$ is a solution of (5) such that $x(t_0, t_0, \bar{x}) = \bar{x}$, then

$$x(t, t_0, \bar{x}) = \bar{x} + \int_{t_0}^{t} f[s, x(s, t_0, \bar{x})]ds \qquad (7)$$

Now if such a solution $x(t, t_0, \bar{x})$ exists and is differentiable with respect to the components $\bar{x}_1, \ldots, \bar{x}_n$ of $\bar{x}$, then we may differentiate both sides of (7) with respect to $\bar{x}_i$ and obtain:

$$\frac{\partial x}{\partial \bar{x}_i} = I_i + \int_{t_0}^{t} \left[\frac{\partial f}{\partial x}[s, x(s, t_0, \bar{x})]\right] \left[\frac{\partial x}{\partial \bar{x}_i}\right] ds \qquad (8)$$

where

$$I_i = \begin{bmatrix} 0 \\ \vdots \\ 1 \\ \vdots \\ 0 \end{bmatrix}$$

i.e., I_i has a 1 in the ith position and zeros elsewhere, and

$$\left[\frac{\partial f}{\partial x}[s, x(s, t_0, \bar{x})]\right]$$

is the $n \times n$ matrix whose entry in the (i, j) position is

$$\frac{\partial f_i}{\partial x_j}[s, x(s, t_0, \bar{x})]$$

This suggests that in order to obtain a solution of (5) which is differentiable with respect to the components of $\bar{x}$, the system of equations described by (7) and (8) should be solved. This is the procedure used.

We set up the system of integral equations

$$x(t) = \bar{x} + \int_{t_0}^{t} f[s, x(s)]ds$$

$$y^{(i)}(t) = I_i + \int_{t_0}^{t} \left[\frac{\partial f}{\partial x}[s, x(s)]\right][y^{(i)}(s)]ds \tag{9}$$

$$(i = 1, \ldots, n)$$

Since f has continuous second partial derivatives with respect to x, it follows that $\frac{\partial f_i}{\partial x_j}(t, x)$ satisfies a Lipschitz condition in x. Solving (9) for $x(t)$ and $y^{(1)}(t), \ldots, y^{(n)}(t)$ is equivalent to solving (7) and (8) for $x(t, t_0, \bar{x})$ and $\frac{\partial x}{\partial \bar{x}_i}$, $i = 1, \ldots, n$.

Let $(x(t, t_0 \bar{x}), y(t, t_0, \bar{x}))$ denote a solution of (9) obtained by applying the successive approximations method used to prove Existence Theorem 1.1. Then $(x(t, t_0, \bar{x}), y^{(1)}(t, t_0, \bar{x}), \ldots, y^{(n)}(t, t_0, \bar{x})$ is the limit (in uniform convergence) on a set

$$\mathcal{R} = \{(t, x)/|t - t_0| \leq a, |x - \bar{x}| \leq b\}$$

of the sequence $\{(x_m(t, t_0 \bar{x}), y_m^{(1)}(t, t_0, \bar{x}), \ldots, y_m^{(n)}(t, t_0, \bar{x})\}$ where

$$(x_0(t, t_0, \bar{x}), y_0^{(1)}(t, t_0, \bar{x}), \ldots, y_0^{(n)}(t, t_0, \bar{x})) = (\bar{x}, 1, \ldots, 1) \tag{10}$$

and

$$x_{m+1}(t, t_0, \bar{x}) = \bar{x} + \int_{t_0}^{t} f[s, x_m(s, t_0, \bar{x})]ds \tag{11}$$

$$y_{m+1}^{(i)}(t, t_0, \bar{x}) = I_i + \int_{t_0}^{t} \left[\frac{\partial f}{\partial x}[s, x_m(s, t_0, \bar{x})]\right][y_m^{(i)}(s, t_0, \bar{x})]ds \tag{12}$$

By (10), we have $y_0^{(i)} = \frac{\partial x_0}{\partial \bar{x}_i}$. Now suppose that for some fixed m and for all $(t, x) \in Int\ \mathcal{R}$ it is true that $\frac{\partial x_m}{\partial \bar{x}_i} = y_m^{(i)}$ for $i = 1, \ldots, n$. Differentiating (11) with respect to $\bar{x}_i$ we obtain:

$$\frac{\partial x_{m+1}}{\partial \bar{x}_i} = I_i + \int_{t_0}^{t} \left[\frac{\partial f}{\partial x}[s, x_m(s, t_0, \bar{x})]\right]\left[\frac{\partial x_m}{\partial \bar{x}_i}(s, t_0, \bar{x})\right]ds$$

$$= I_i + \int_{t_0}^{t} \left[\frac{\partial f}{\partial x}[s, x_m(s, t_0, \bar{x})]\right]\left[y_m^{(i)}(s, t_0, \bar{x})\right]ds$$

$$= y_m^{(i)}(t, t_0, \bar{x})$$

Thus by induction it follows that for all m and for all $(t,x) \in Int\ \mathcal{R}$ it is true that $\frac{\partial x_m}{\partial \bar{x}_i} = y_m^{(i)}$. But the sequence

$$\{y_m^{(i)}(t,t_0,\bar{x})\} = \{\frac{\partial x_m}{\partial \bar{x}_i}(t,t_0,\bar{x})\}$$

converges uniformly on $\mathcal{R}$ to $y^{(i)}(t,t_0,\bar{x})$. Hence by a standard convergence theorem from calculus, it follows that

$$y^{(i)}(t,t_0,\bar{x}) = \frac{\partial x}{\partial \bar{x}_i}(t,t_0,\bar{x}), \quad i = 1,\ldots,n$$

This completes the proof of the Differentiability Theorem. $\square$

Existence Theorem for Equation With a Parameter

It is natural to suppose that if $f(t,x,\mu)$ is continuous on (t,x)-space and is also continuous in a parameter μ, then solutions of the equation

$$x' = f(t,x,\mu)$$

also depend continuously on the parameter μ. For later work, we will need a result of this kind.

Existence Theorem for Equation with a Parameter *Consider the equation*

$$\frac{dx}{dt} = f(t,x,\mu) \tag{13}$$

and suppose that f is defined and continuous on

$$(a,b) \times \bar{G} \times [-\tilde{\mu}, \tilde{\mu}]$$

where $a < b$, and $\bar{G}$ is the closure of a bounded open set G in R^n, and $\tilde{\mu} > 0$. Suppose that the function $f(t,x,\mu)$ satisfies a Lipschitz condition with respect to x in $(a,b) \times \bar{G} \times [-\tilde{\mu}, \tilde{\mu}]$ where the Lipschitz constant is independent of μ. Let $t_0 \in (a,b)$ and let $x(t,x^0,\mu)$ be the solution of (13) such that

$$x(t_0,x^0,\mu) = x^0$$

where $x_0 \in G$. Then $x(t, x^0, \mu)$ is a continuous function of (x^0, μ) and there exists $r > 0$ such that $x(t, x^0, \mu)$ is continuous in (x^0, μ) uniformly for $t \in [t_0 - r, t_0 + r]$.

Proof. Let $\mu_1, \mu_2 \in [-\tilde{\mu}, \tilde{\mu}]$ and suppose $x_1^0, x_2^0 \in G$. By Lemma 1.1, we have: if $|t - t_0| \leq r$ where r is a sufficiently small positive number, then

$$
\begin{aligned}
|x(t, x_2^0, \mu_2) - x(t, x_1^0, \mu_1)| = &|x_2^0 - x_1^0 + \int_{t_0}^t \{f[s, x(s, x_2^0, \mu_2), \mu_2] \\
&- f[s, x(s, x_1^0, \mu_1), \mu_1]\}ds| \\
= &|x_2^0 - x_1^0 + \int_{t_0}^t \{f[s, x(s, x_2^0, \mu_2), \mu_2] \\
&- f[s, x(s, x_1^0, \mu_1), \mu_2] \\
&+ f[s, x(s, x_1^0, \mu_1), \mu_2] \\
&- f[s, x(s, x_1^0, \mu_1), \mu_1]\}ds| \\
\leq &|x_2^0 - x_1^0| \\
&+ k|t - t_0| \max_{|t-t_0| \leq r} |x(t, x_2^0, \mu_2) - x(t, x_1^0, \mu_1)| \\
&+ |t - t_0| \max_{|t-t_0| \leq r} |f[t, x(t, x_1^0, \mu_1), \mu_2] \\
&- f[t, x(t, x_1^0, \mu_1), \mu_1]|
\end{aligned}
\tag{14}
$$

where k is the Lipschitz constant given by hypothesis. Now we choose r small enough so that

$$
k|t - t_0| \leq \frac{1}{2}.
\tag{15}
$$

Applying (15) to (14), we obtain

$$
\begin{aligned}
\max_{|t-t_0| \leq r} |x(t, x_2^0, \mu_2) - x(t, x_1^0, \mu_1)| \leq &|x_2^0 - x_1^0| \\
&+ \frac{1}{2} \max_{|t-t_0| \leq r} |x(t, x_2^0, \mu_2) - x(t, x_1^0, \mu_1)| \\
&+ r \max_{|t-t_0| \leq r} |f[t, x(t, x_1^0, \mu_1), \mu_2] - f[t, x(t, x_1^0, \mu_1), \mu_1]|
\end{aligned}
$$

and hence

$$
\max_{|t-t_0| \leq r} |x(t, x_2^0, \mu_2) - x(t, x_1^0, \mu_1)|
$$

$$\leq 2|x_2^0 - x_1^0| + 2r \max_{|t-t_0|\leq r} |f[t, x(t, x_1^0, \mu_1), \mu_2]$$

$$-f[t, x(t, x_1^0, \mu_1), \mu_1]| \tag{16}$$

But $f[t, x(t, x_1^0, \mu_1), \mu]$ is uniformly continuous on

$$\{t/|t - t_0| \leq r\} \times \{\mu/ - \tilde{\mu} \leq \mu \leq \tilde{\mu}\}.$$

Hence if $|x_2^0 - x_1^0|$ and $|\mu_2 - \mu_1|$ are sufficiently small it follows from (16) that

$$\max_{|t-t_0|\leq r} |x(t, x_2^0, \mu_2) - x(t, x_1^0, \mu_1)|$$

can be made less than ε. Thus we have shown that $x(t, x^0, \mu)$ is continuous at (x_1^0, μ_1) uniformly for $|t - t_0| \leq r$. $\square$

Note 1: The proof shows that the hypothesis concerning the Lipschitz condition of f can be weakened to a local Lipschitz condition. Stating the weaker condition is messy and this messiness outweighs the theoretical advantage.

Note 2: For an extended version of the Existence Theorem for an Equation with a Parameter, see Exercise 13.

Existence Theorem Proved by Using a Contraction Mapping

The technique used in the proof of Existence Theorem 1.1, the method of successive approximations, is used in many parts of analysis, and it is used in a generalized form to prove a standard existence theorem in functional analysis, the Banach Fixed Point Theorem or Principle of Contraction Mappings (see Appendix).

Our next step is to prove a slightly different version of the basic existence theorem by using the Banach Fixed Point Theorem. Using a result from functional analysis makes the proof simpler and shorter. After we obtain our second version of the existence theorem we will make a more extensive comparison of it with Existence Theorem 1.1.

As before, let D be an open set in (t, x)-space and suppose that $(t_0, x^0) \in D$. Let A, B be positive numbers such that the set

$$\mathcal{D} = \{(t, x)/|t - t_0| \leq A, |x - x^0| \leq B\}$$

is contained in D. Let

$$M = \max_{(t,x)\in\mathcal{D}} |f(t,x)|$$

Let a,b be positive numbers which are small enough so that if $|x^1 - x^0| \leq b$, then the "cone"

$$\mathcal{K}_{x^1} = \{(t,x)/|t - t_0| \leq a, |x - x^1| \leq M|t - t_0|\}$$

is contained in $\mathcal{D}$. See Figure 2 where the constants a, b are such that $Ma \leq B - b$ and the set $\mathcal{K}_{x^1}$, where $x^1 - x^0 = b$, is shaded vertically.

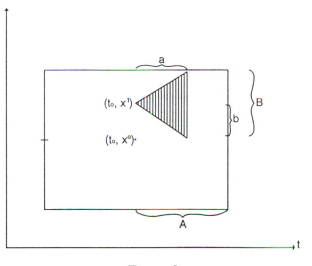

Figure 2

Let $\mathcal{M}$ be the space of continuous mappings m from the domain

$$\mathcal{R} = \{(t,x)/|t - t_0| \leq a; |x - x^0| \leq b\}$$

into R^n and let $\mathcal{E}$ be the set

$$\mathcal{E} = \{m \in \mathcal{M}/|m(t,x)| \leq M|t - t_0|\}$$

It is easy to show that the function ρ, with domain $\mathcal{E} \times \mathcal{E}$, defined by

$$\rho(m_1, m_2) = \max_{(t,x)\in\mathcal{R}} |m_1(t,x) - m_2(t,x)|$$

where $m_1, m_2 \in \mathcal{E}$, is a metric on $\mathcal{E}$. By standard theorems from calculus, it follows that $\mathcal{E}$ is a complete metric space.

Now let $\mathcal{S}$ be the mapping from domain $\mathcal{E}$ into $\mathcal{M}$ such that if $m \in \mathcal{E}$, then $\mathcal{S}m$ is defined by:

$$(\mathcal{S}m)(t, x) = \int_{t_0}^{t} f[s, x + m(s, x)]ds \qquad (17)$$

We suppose that f satisfies a Lipschitz condition (with Lipschitz constant k) on $\mathcal{R}$.

Lemma 1.6 *If the positive number a is sufficiently small, mapping $\mathcal{S}$ is a contraction mapping from $\mathcal{E}$ into $\mathcal{E}$. That is, there exists a real number $q \in (0, 1)$ such that if $m_1, m_2 \in \mathcal{E}$, the $\mathcal{S}m_1, \mathcal{S}m_2 \in \mathcal{E}$ and*

$$\rho(\mathcal{S}m_1, \mathcal{S}m_2) \leq q[\rho(m_1, m_2)]. \qquad (18)$$

Proof. We must show:

1. $\mathcal{S}$ is well-defined, i.e., the point $(s, x + m(s, x))$ is in the domain of mapping f so that the right-hand of (17) makes sense;

2. $\mathcal{S}$ takes $\mathcal{E}$ into $\mathcal{E}$, i.e., if $m \in \mathcal{E}$, then $\mathcal{S}m \in \mathcal{E}$;

3. If a is sufficiently small, then $\mathcal{S}$ satisfies inequality (18)

Proof of (i). Let $(s, x) \in \mathcal{R}$ and let

$$K_x = \{(t, y)/|t - t_0| \leq a, |y - x| \leq M|t - t_0|\}$$

Then $(s, x + m(s, x))$ is contained in K_x because since $(s, x) \in \mathcal{R}$, then

$$|s - t_0| \leq a$$

and

$$|[x + m(s, x)] - x| = |m(s, x)| \leq M|s - t_0|$$

But $K_x \subset \mathcal{D}$ and $\mathcal{D}$ is a subset of the domain of f.

Proof of (ii). By standard theorems from calculus, $\mathcal{S}m$ is continuous on $\mathcal{R}$ and

$$|(\mathcal{S}m)(t, x)| \leq M|t - t_0|$$

Proof of (iii). Since f satisfies a Lipschitz condition with Lipschitz constant k on $\mathcal{R}$, then

$$|(\mathcal{S}m_1)(t,x) - (\mathcal{S}m_2)(t,x)|$$

$$\leq \int_{t_0}^t |f[s, x + m_1(s,x)] - f[s, x + m_2(s,x)]|ds$$

$$\leq k \int_{t_0}^t |m_1(s,x) - m_2(s,x)|ds$$

$$\leq k[\rho(m_1, m_2)]|t - t_0|$$

$$\leq ka\rho(m_1, m_2)$$

Thus

$$\rho[\mathcal{S}(m_1) - \mathcal{S}(m_2)] \leq ka\rho(m_1, m_2)$$

and if $ka < 1$, inequality (18) is satisfied. This completes the proof of Lemma 1.6. □

Remark. Since k is a Lipschitz constant for f on the set $\mathcal{R}$, then if we choose a small enough so that $ka < 1$, the set $\mathcal{R}$ is either unchanged or made smaller. Hence k is still a Lipschitz constant for f on the (possibly smaller) set $\mathcal{R}$.

Lemma 1.6 permits us to apply the Banach Fixed Point Theorem or Principle of Contraction Mappings because Lemma 1.6 shows that $\mathcal{S}$ is a contraction mapping from the complete metric space $\mathcal{M}$ into itself. The Banach Fixed Point Theorem shows that mapping $\mathcal{S}$ has a unique fixed point $\bar{m}(t,x) \in \mathcal{E}$, i.e.,

$$\bar{m}(t,x) = \int_{t_0}^t f[s, x + \bar{m}(s,x)]ds \tag{19}$$

Let $x = \bar{x}$, where $|\bar{x} - x^0| \leq b$, be fixed in (19) and add $\bar{x}$ to each side of (19). We obtain:

$$\bar{x} + \bar{m}(t,\bar{x}) = \bar{x} + \int_{t_0}^t f[s, \bar{x} + \bar{m}(s,\bar{x})]ds \tag{20}$$

Differentiation with respect to t on both sides of (20) shows that $\bar{x} + \bar{m}(t,x)$ is a solution $x(t)$ of (1) such that

$$x(t_0) = \bar{x} \tag{21}$$

Moreover $\bar{x} + \bar{m}(t, \bar{x})$ is a continuous function of $\bar{x}$.

By Lemma 1.1,

$$x(t) = \bar{x} + \int_{t_0}^{t} f[s, x(s)] ds$$

and hence

$$|x(t) - \bar{x}| \leq M|t - t_0|$$

Since $\bar{m}(t, x)$ is a unique fixed point in $\mathcal{E}$ of $\mathcal{S}$, it follows that $x(t)$ is a unique solution of (1) satisfying the initial condition (21). Hence we obtain the following somewhat extended version of our existence theorem.

Existence Theorem 1.2 *Let D be an open set in (t, x)-space. Let $(t_0, x^0) \in D$, and suppose function f is defined and continuous on D and satisfies a Lipschitz condition with respect to x on D. Then there exist positive numbers a, b such that if $|\bar{x} - x^0| \leq b$, the differential equation*

$$x' = f(t, x)$$

has a unique solution $x(t, t_0, \bar{x})$ on $(t_0 - a, t_0 + a)$ such that

$$x(t_0, t_0, \bar{x}) = \bar{x}$$

and this solution is a continuous function of $(t, \bar{x})$.

In comparing Existence Theorems 1.1 and 1.2, we notice first a fundamental similarity: both are proved by using successive approximations. Successive approximations are used directly in the proof of Existence Theorem 1.1. In Existence Theorem 1.2, the proof is obtained by applying the Banach Fixed Point Theorem, and the Banach Theorem, in turn, is proved by using successive approximations.

Existence Theorem 1.1 has, of course, the advantage that its proof is straightforward and used only results from calculus. Its second advantage is that the domain of the solution, i.e., the interval $(t_0 - A, t_0 + A)$, is easily computed ($A = \min(a, b/M)$) whereas the domain of the solution obtained in Existence Theorem 1.2 is not described as explicitly and, moreover, it may be smaller than the domain obtained in Existence Theorem 1.1. For Existence Theorem 1.1, the numbers a and b simply indicate the dimensions of a rectangle "centered about" (t_0, x^0) and contained in D whereas for Existence Theorem 1.2, the numbers a and b have to be small enough so that the "cone" K_x is contained in $\mathcal{D}$ (as shown in Figure 2). The numbers a and b have to be calculated in each case. Also, as Figure 2 shows, the numbers a and b may be considerably smaller than the numbers a and b used in Existence Theorem 1.1. Finally, as shown in the proof of Lemma 1.6, the number a used in Existence Theorem 1.2 must be smaller than the reciprocal of the Lipschitz constant of f. The numbers a and b used in Existence Theorem 1.1 are completely independent of the Lipschitz constant of f. These advantages of Existence Theorem 1.1 are more apparent than real because, as pointed out earlier, it is not important at this stage to try to maximize the interval of existence of the solution. (We take up this question in the Extension Theorems later in this chapter.) Existence Theorem 1.2 has, on the other hand, two significant advantages. Its proof uses a result from functional analysis (the Banach Fixed Point Theorem) but the very use of this abstract theorem relates the proof of Existence Theorem 1.2 to other parts of analysis where the Banach Fixed Point Theorem is used. Another advantage of Existence Theorem 1.2 is that part of the conclusion is the fact that the solution depends continuously on the initial condition x^0 whereas with Existence Theorem 1.1 we gave a separate proof of continuity (Corollary 1.1).

Existence Theorem Without Uniqueness

We used the Lipschitz condition of f heavily in the proofs of both Existence Theorem 1.1 and Existence Theorem 1.2. We needed the Lipschitz condition to prove both existence and uniqueness. The next question we consider is the following: is it possible to reduce the hypothesis on f? As we will see, if the Lipschitz condition on f is replaced by a mere continuity condition on f, then we can still prove the existence of a solution but we lose the uniqueness. First we describe a simple example which shows that if f does not satisfy a Lipschitz condition, then the solution may not be unique. Let x

be a scalar (or 1-vector) and consider the equation

$$\frac{dx}{dt} = x^{1/3} \tag{22}$$

Let $k \in (0,1)$ and define the function $x_k(t)$ as follows:

$$x_k(t) = 0 \qquad for\ t \in (0,k]$$

$$x_k(t) = \left[\frac{2}{3}(t-k)\right]^{3/2} \qquad for\ t \in (k,1)$$

If $t_0 \in (k,1)$ then the derivative of $x_k(t)$ at t_0 is:

$$\frac{dx_k}{dt} = \frac{3}{2}\left[\frac{2}{3}(t_0-k)\right]^{1/2}\left(\frac{2}{3}\right) = [x_k(t_0)]^{1/3}$$

If $t_0 \in (0,k)$, the derivative at t_0 is:

$$\frac{dx_k}{dt} = 0 = [x_k(t_0)]^{1/3}$$

If $t_0 = k$, the left-hand derivative is clearly 0, and it is a short calculation to show that the right-hand derivative is zero. Hence $x_k(t)$ is a solution of (22). But for fixed $t_0 \in (0,1)$, the initial condition $x(t_0) = 0$ is satisfied by each $x_k(t)$ with $k \geq t_0$.

   For an example in which there is more than one solution for every initial condition, see Hartman [1964, page 18].

   We are left with the problem of proving the existence of a solution if f is merely continuous, i.e., if f does not satisfy a Lipschitz condition. We will prove this existence essentially by using techniques from calculus. There are several reasons for proving such an existence result. First, from the pure mathematics viewpoint, we want to obtain as clear a picture as possible of what conditions are needed to insure existence. Second the technique to be used in the proof is the basis for a method often used in numerical analysis. Finally, the theorem we will prove makes possible a quick approach to the problem of estimating how large the domain of the solution is.

Existence Theorem 1.3 *Let $\bar{G}$ be the closure of a bounded open set G in R^n and let $f_1, \ldots, f_n$ be real-valued and continuous on $\bar{G}$. Let t_0 be a fixed real number and let*

$$x^0 = (x_1^0, \ldots, x_n^0)$$

be a fixed point in G. Then there exist functions $x_1(t), \ldots, x_n(t)$ satisfying the following conditions:

1. *Each $x_i(t)$, $i = 1, \ldots, n$, is defined on a domain which contains the interval*

$$I = \left[t_0 - \frac{d}{M\sqrt{n}}, t_0 + \frac{d}{M\sqrt{n}} \right]$$

   where $d = \inf_{q \in \bar{G} - G} \|x^0 - q\|$ and $\|x^0 - q\|$ denotes the Euclidean norm and M is an upper bound for the set

$$\{|f_i(x_1, \ldots, x_n)| / (x_1, \ldots, x_n) \in \bar{G}, i = 1, \ldots, n\}$$

   (Since $x^0 \in G$, then $d > 0$. Since $\bar{G}$ is bounded, M is finite.)

2. *$(x_1(t_0), \ldots, x_n(t_0)) = x^0$.*

3. *$(x_1(t), \ldots, x_n(t))$ is a solution on $(t_0 - \frac{d}{M\sqrt{n}}, t_0 + \frac{d}{M\sqrt{n}})$ of the system*

$$x'_i = f_i(x_1, \ldots, x_n), \quad i = 1, \ldots, n \tag{23}$$

Remark. Notice first that Existence Theorem 1.3 applies only to autonomous equations, i.e., systems in which the functions $f_i(1 = i, \ldots, n)$ are independent of t. However, this limitation is only apparent because the problem of solving a nonautonomous system can be reduced to the problem of solving an autonomous system by using the following procedure. Suppose we seek a solution $(x_1(t), \ldots, x_n(t))$ of the system

$$x'_i = g_i(t, x_1, \ldots, x_n), \quad i = 1, \ldots, n \tag{24}$$

such that

$$x_i(t_0) = x_i^0 \quad i = 1, \ldots, n \tag{25}$$

where the point $(t_0, x_1^0, \ldots, x_n^0)$ is in the interior of the domain of the functions g_i $(i = 1, \ldots, n)$. Consider the autonomous system:

$$\frac{dx_1}{d\tau} = g_1(t, x_1, \ldots, x_n)$$

$$\cdots$$

$$\frac{dx_n}{d\tau} = g_n(t, x_1, \ldots, x_n) \tag{26}$$

$$\frac{dt}{d\tau} = 1$$

Suppose that system (26) has a solution $(\bar{x}_1(\tau), \ldots, \bar{x}_n(\tau), t(\tau))$ such that

$$x_i(t_0) = x_i^0 \quad i = 1, \ldots, n$$
$$t(t_0) = t_0$$

Since $dt/d\tau = 1$, then $t(\tau) = \tau + c$ where c is a constant. Since $t(t_0) = t_0$, then $c = 0$ and $t(\tau) = \tau$. Thus $(\bar{x}_1(t), \ldots, \bar{x}_n(t))$ is a solution of (24) which satisfies the initial condition (25).

Proof of Existence Theorem 1.3. The method of proof is called the Euler or Cauchy-Euler method, and it is the basis for several methods in numerical analysis. First we explain the underlying idea of the method for the 2-dimensional case.

The set $\bar{G}$ is "chopped up" into small rectangles and subsets of rectangles by using lines parallel to the x_1-axis and the x_2-axis. See Figure 3.

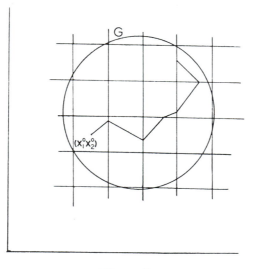

Figure 3

Draw the line segment which starts at the point (x_1^0, x_2^0), which is in one of the rectangles, and has slope

$$\frac{f_2(x_1^0, x_2^0)}{f_1(x_1^0, x_2^0)}$$

and extend the line segment until it reaches the boundary of the rectangle, say at the point (x_1^1, x_2^1). Take the line segment which starts at (x_1^1, x_2^1) and has slope

$$\frac{f_2(x_1^1, x_2^1)}{f_1(x_1^1, x_2^1)}.$$

and extend the line segment until it reaches the boundary of the rectangle. Continue in this way and obtain a broken line as shown in Figure 3. Corresponding to each "chopping up" of $\bar{G}$ by lines parallel to the x_1-axis and the x_2-axis such a broken line is obtained. As smaller and smaller rectangles are used, the broken lines obtained become better and better approximations to a solution of the system

$$x_1' = f_1(x_1, x_2)$$
$$x_2' = f_2(x_1, x_2)$$

The proof of Existence Theorem 1.3 consists in giving a rigorous, n-dimensional account of the procedure just described. It is interesting to notice that whereas the underlying idea of the method can be described almost instantaneously by using Figure 3, a detailed rigorous account, which follows, makes burdensome reading. This is a good example of work that often occurs in mathematics: the decision to use a simple but powerful and far-reaching idea is followed by the onerous labor of showing that the idea can actually be used.

First, since G is bounded, then $\bar{G}$ is compact. Hence the functions $f_1, \ldots, f_n$ are uniformly continuous on $\bar{G}$. That is, if $\varepsilon > 0$, then there exists $\delta > 0$ such that if

$$\sum_{i=1}^{n} \left| x_i^{(1)} - x_i^{(2)} \right| < \delta$$

then for $i = 1, \ldots, n$,

$$\left| f_i(x_1^{(1)}, \ldots, x_n^{(1)}) - f_i(x_1^{(2)}, \ldots, x_n^{(2)}) \right| < \varepsilon$$

Take a fixed $\varepsilon > 0$ and by using planes parallel to the coordinate planes in R^n, divide $\bar{G}$ into closed cubes of side with length δ/n and subsets of these cubes. Denote these cubes and subsets of cubes by $K_1, \ldots, K_s$. Choose the planes parallel to the coordinate planes in R^n so that x^0 is in the interior of some K_i, say K_1. Define the functions

$$x_i^{(1)}(t) = x_i^0 + [f_i(x_1^0, \ldots, x_n^0)](t - t_0)$$

where $i = 1, \ldots, n$ and $t \in [t_0, t_1]$ and

$$t_1 = lub\{t/t \geq t_0, (x_1^{(1)}(\tau), \ldots, x_n^{(1)}(\tau)) \in K_1 \ for \ \tau \in [t_0, t]\}$$

(Notice that $t_1 > t_0$ because x^0 is in the interior of K_1.) If $t_1 = \infty$, then since K_1 is bounded, it follows that for $i = 1, \ldots, n$

$$f_i(x_1^0, \ldots, x_n^0) = 0$$

Thus if $t_1 = \infty$, the solution promised in the statement of the theorem is: for $i = 1, \ldots, n$

$$x_i(t) = x_i^0$$

for all real t. If $t_1 < \infty$, let

$$x_i^{(1)} = \lim_{t \uparrow t_1} x_i^{(1)}(t) = x_i^0 + [f_i(x_1^0, \ldots, x_n^0)](t_1 - t_0)$$

If $\left(x_1^{(1)}, \ldots, x_n^{(1)}\right) \in \bar{G} - G$, proceed no further. If $\left(x_1^{(1)}, \ldots, x_n^{(1)}\right) \in G$, define:

$$x_i^{(2)}(t) = x_i^{(1)} + [f_i\left(x_1^{(1)}, \ldots, x_n^{(1)}\right)](t - t_1) \ (i = 1, \ldots, n)$$

Then there exists $t^{(1)} > t_1$ and a K_2 (K_2 may be K_1) such that if $\tau \in [t_1, t^{(1)}]$, then

$$\left(x_1^{(2)}(\tau), \ldots, x_n^{(2)}(\tau)\right) \in K_2$$

Let

$$t_2 = lub\{t/t \geq t_1, \left(x_1^{(2)}(\tau), \ldots, x_n^{(2)}(\tau)\right) \in K_2 \ for \ \tau \in [t_1, t]\}$$

and let

$$x_i^{(2)} = \lim_{t \uparrow t_2} x_i^{(2)}(t) = x_i^{(1)} + \left[f_i\left(x_1^{(1)}, \ldots, x_n^{(1)}\right) \right](t_2 - t_1)$$

If

$$f_1\left(x_1^{(1)}, \ldots, x_n^{(1)}\right) = \cdots = f_n\left(x_1^{(1)}, \ldots, x_n^{(1)}\right) = 0$$

then $t_2 = \infty$ and for all $t \geq t_1$,

$$x_i^{(2)}(t) = x_i^{(1)} \quad (i = 1, \ldots, n)$$

We proceed by induction to define

$$x_i^{(k+1)}(t) = x_i^{(k)} + [f_i(x_1^{(k)}, \ldots, x_n^{(k)})](t - t_k)$$

with domain $[t_k, t_{k+1}]$. (The induction process comes to an end after k steps if

$$(x_1^{(k)}, \ldots, x_n^{(k)}) \in \bar{G} - G$$

or if $t_k = \infty$.)

Next we "piece together" the domains $[t_0, t_1], [t_1, t_2], \ldots$ and the vector functions $(x_1^{(1)}(t), \ldots, x_n^{(1)}(t)), (x_1^{(2)}(t), \ldots, x_n^{(2)}(t)), \ldots$ and obtain a new vector function $(\bar{x}_1(t), \ldots, \bar{x}_n(t))$ with domain

$$\bigcup_{i \geq 0} [t_i, t_{i+1}]$$

and such that if $t \in [t_i, t_{i+1}]$, then

$$\bar{x}_j(t) = x_j^{(i+1)}(t)$$

for $j = 1, \ldots, n$. From the definition, it follows that each $\bar{x}_j(t)$ is continuous on

$$\bigcup_{i \geq 0} [t_i, t_{i+1}].$$

A sufficient condition that $(x_1^{(s)}, \ldots, x_n^{(s)}) \in \bar{G}$ is that

$$\sum_{j=0}^{s-1} \left\{ \sum_{i=1}^{n} [f_i(x_1^{(j)}, \ldots, x_n^{(j)})]^2 (t_{j+1} - t_j)^2 \right\}^{1/2} \leq d \qquad (27)$$

Since

$$\left\{ \sum_{i=1}^{n} [f_i(x_1^{(j)}, \ldots, x_n^{(j)})]^2 (t_{j+1} - t_j)^2 \right\}^{1/2} \leq \{nM^2(t_{j+1} - t_j)^2\}^{1/2}$$

$$\leq M\sqrt{n}(t_{j+1} - t_j)$$

then a sufficient condition that (27) hold is:

$$\sum_{j=0}^{s-1} M\sqrt{n}(t_{j+1} - t_j) \leq d$$

$$M\sqrt{n}(t_s - t_0) \leq d$$

$$(t_s - t_0) \leq \frac{d}{M\sqrt{n}}$$

Thus the domain of $(\bar{x}_1(t), \ldots, \bar{x}_n(t))$ contains the interval

$$\left[t_0, t_0 + \frac{d}{M\sqrt{n}} \right]$$

We proceed in an exactly similar way to consider $t < t_0$ and obtain finally the n-vector function $(\bar{x}_1(t), \ldots, \bar{x}_n(t))$ whose domain contains the interval

$$I = \left[t_0 - \frac{d}{M\sqrt{n}}, t_0 + \frac{d}{M\sqrt{n}} \right]$$

Lemma 1.7 *For $i = 1, \ldots, n$ and $t \in I$,*

$$\bar{x}_i(t) = x_i^0 + \int_{t_0}^{t} f_i [\bar{x}_1(s), \ldots, \bar{x}_n(s)] \, ds + \int_{t_0}^{t} g_i(s) ds \qquad (28)$$

where the function g_i is such that for all $s \in I$,

$$|g_i(s)| < \varepsilon$$

Proof.

$$x_i^{(1)} - x_i^0 = \int_{t_0}^{t_1} f_i \left(x_1^0, \ldots, x_n^0 \right) ds$$

$$x_i^{(2)} - x_i^{(1)} = \int_{t_1}^{t_2} f_i \left(x_1^{(1)}, \ldots, x_n^{(1)} \right) ds$$

$$\cdots$$

$$x_i^{(k)} - x_i^{(k-1)} = \int_{t_{k-1}}^{t_k} f_i(x_1^{(k-1)}, \ldots, x_n^{(k-1)}) ds$$

Adding these equations together we obtain:

$$x_i^{(k)} - x_i^0 = \sum_{j=0}^{k-1} \int_{t_j}^{t_{j+1}} f_i(x_1^{(j)}, \ldots, x_n^{(j)}) ds \qquad (29)$$

(where $x_i^{(0)} = x_i^0$ for $i = 1, \ldots, n$).
  If $t_k \leq t \leq t_{k+1}$,

$$\int_{t_k}^{t} f_i(x_1^{(k)}, \ldots, x_n^{(k)}) ds = [f_i(x_1^{(k)}, \ldots, x_n^{(k)})](t - t_k)$$
$$= x_i^{(k+1)}(t) - x_i^{(k)} = \bar{x}_i(t) - x_i^{(k)} \qquad (30)$$

From equations (29) and (30), we obtain: if $t_k \leq t \leq t_{k+1}$,

$$\bar{x}_i(t) = x_i^{(k)} + \int_{t_k}^{t} f_i(x_1^{(k)}, \ldots, x_n^{(k)}) ds$$

$$= x_i^0 + \sum_{j=0}^{k-1} \int_{t_j}^{t_{j+1}} f_i(x_1^{(j)}, \ldots, x_n^{(j)}) ds$$

$$+ \int_{t_k}^{t} f_i(x_1^{(k)}, \ldots, x_n^{(k)}) ds$$

Hence if $t_k \leq t \leq t_{k+1}$,

$$\bar{x}_i(t) - x_i^0 = \sum_{j=0}^{k-1} \int_{t_j}^{t_{j+1}} f_i[x_1^{(j)}(s), \ldots, x_n^{(j)}(s)] ds$$

$$+ \int_{t_k}^{t} f_i[x_1^{(k)}(s), \ldots, x_n^{(k)}(s)] ds$$

$$+ \sum_{j=0}^{k-1} \int_{t_j}^{t_{j+1}} \{f_i(x_1^{(j)}, \ldots, x_n^{(j)}) - f_i(x_1^{(j)}(s), \ldots, x_n^{(j)}(s))\} ds$$

$$+ \int_{t_k}^{t} \{f_i(x_1^{(k)}, \ldots, x_n^{(k)}) - f_i[x_1^{(k)}(s), \ldots, x_n^{(k)}(s)]\} ds$$

$$= \int_{t_0}^{t} f_i[\bar{x}_1(s), \ldots, \bar{x}_n(s)] ds + \int_{t_0}^{t} g_i(s) ds$$

where if $s \in [t_j, t_{j+1}]$, $j = 1, \ldots, k - 1$

$$|g_i(s)| = |f_i(x_1^{(j)}, \ldots, x_n^{(j)}) - f_i(x_1^{(j)}(s), \ldots, x_n^{(j)}(s))| < \varepsilon$$

because $(x_1^{(j)}, \ldots x_n^{(j)})$ and $(x_1^{(j)}(s), \ldots, x_n^{(j)}(s))$ are in the same set K_j. A similar estimate holds for $|g_i(s)|$ if $s \in [t_k, t]$. This completes the proof of Lemma 1.7.   □

The n-vector function $(\bar{x}_1(t), \ldots, \bar{x}_n(t))$ was obtained corresponding to a number $\varepsilon > 0$. Next we take $\varepsilon = 1/m$, where m is a positive integer, and let $(\bar{x}_1^m(t), \ldots, \bar{x}_n^m(t))$ be the corresponding n-vector function. It follows from (28) that if $t \in I$,

$$|\bar{x}_i^m(t)| \le B + M|t - t_0| + \frac{1}{m}|t - t_0|$$

where $B = \max_{i=1,\ldots,n} |x_i^0|$. It also follows from (28) that if $t', t'' \in I$, then

$$|\bar{x}_i^m(t') - \bar{x}_i^m(t'')| \le M|t' - t''| + \frac{1}{m}|t' - t''|$$

Thus $\{\bar{x}_i^m(t)\}$ is a sequence of uniformly bounded, equicontinuous functions on I. Hence by Ascoli's Theorem, there is a subsequence $\{(\bar{x}_1^{m_\ell}(t), \ldots, \bar{x}_n^{m_\ell}(t))\}$ such that for $i = 1, \ldots, n$ $\{\bar{x}_i^{m_\ell}(t)\}$ converges uniformly on I to a continuous function $u_i(t)$. Now we have:

$$\bar{x}_i^{m_\ell}(t) = x_i^0 + \int_{t_0}^t f_i\left[\bar{x}_1^{m_\ell}(s), \ldots, \bar{x}_n^{m_\ell}(s)\right] ds + \int_{t_0}^t g_i^{m_\ell}(s) ds \qquad (31)$$

If $t \in I$,

$$\left|\int_{t_0}^t g_i^{m_\ell}(s) ds\right| \le \frac{1}{m_\ell}|t - t_0| \le \frac{1}{m_\ell} \frac{d}{\sqrt{n}\, M}$$

Hence $\int_{t_0}^t g_i^{m_\ell}(s) ds$ converges uniformly to zero on I. Also we have

$$\left|\int_{t_0}^t f_i\left[\bar{x}_i^{m_\ell}(s), \ldots, \bar{x}_n^{m_\ell}(s)\right] ds - \int_{t_0}^t f_i[u_1(s), \ldots, u_n(s)] ds\right|$$

$$\le |t - t_0| \max_{s \in [t_0, t]} |f_i\left[\bar{x}_i^{m_\ell}(s), \ldots, \bar{x}_n^{m_\ell}(s)\right] - f_i[u_1(s), \ldots, u_n(s)]|$$

$$< |t - t_0|\varepsilon$$

The last inequality holds for m_ℓ sufficiently large because for $i = 1, \ldots, n, \bar{x}_i^{m_\ell}(t)$ converges uniformly to $u_i(t)$ and f_i is uniformly continuous on $\bar{G}$. Thus the function

$$\int_{t_0}^t f_i\left[\bar{x}_1^{m_\ell}(s), \ldots, \bar{x}_n^{m_\ell}(s)\right] ds$$

converges uniformly on I to:

$$\int_{t_0}^{t} f_i[u_1(s), \ldots, u_n(s)] ds$$

Thus from (31) it follows that for $t \in I = [t_0 - d/M\sqrt{n}, \, t_0 + d/M\sqrt{n}]$

$$u_i(t) = x_i^0 + \int_{t_0}^{t} f_i[u_1(s), \ldots, u_n(s)] ds \qquad (32)$$

Since $u_1(s), \ldots, u_n(s)$ are continuous and f is continuous, then we may differentiate the right-hand side of (32) and conclude that for each $t \in (t_0 - d/M\sqrt{n}, \, t_0 + d/M\sqrt{n})$,

$$\frac{du_i}{dt} = f_i[u_1(t), \ldots, u_n(t)], \, i = 1, \ldots, n$$

This completes the proof of Existence Theorem 1.3. $\qquad \square$

Notice that the proof of Existence Theorem 1.3 is not constructive because when Ascoli's Theorem is applied, we can only conclude that there exists a subsequence $\{(\bar{x}_1^{m_\ell}(t), \ldots, \bar{x}_n^{m_\ell}(t))\}$ which converges to a solution. We do not exhibit the subsequence. However, if enough hypotheses are imposed so that the solution of (23) which satisfies the initial value is unique, then as the following corollary shows, the sequence $\{(\bar{x}_1(t), \ldots, \bar{x}_n(t))\}$ itself converges to the solution.

Corollary to Existence Theorem 1.3 *Suppose that equation* (23) *is such that if a solution satisfying the initial condition exists, it is unique (a sufficient condition for this is that f satisfies a Lipschitz condition). Then the sequence $\{(\bar{x}_1^m(t), \ldots, \bar{x}_n^m(t))\}$ obtained in the proof of Existence Theorem 1.3 converges to a solution of* (23).

Proof. The proof of Existence Theorem 1.3 shows that there is a subsequence of $\{(\bar{x}_1^m(t), \ldots, \bar{x}_n^m(t))\}$ which converges uniformly on interval I to a solution $(u_1(t), \ldots, u_n(t))$ of (23). Suppose that $\{(\bar{x}_1^m(t), \ldots, \bar{x}_n^m(t))\}$ itself does not converge uniformly to $(u_1(t), \ldots, u_n(t))$ on the interval I. Then there is an $\varepsilon_1 > 0$ and a subsequence

$$\{(\bar{x}_1^{m_v}(t), \ldots, \bar{x}_n^{m_v}(t))\}$$

of $\{(\bar{x}_1^m(t), \ldots, \bar{x}_n^m(t))\}$ such that for each m_v there is a number $t_v \in I$ such that

$$\sum_{i=1}^{n} |u_i(t_v) - \bar{x}_i^{m_v}(t_v)| > \varepsilon_1 \tag{33}$$

But the proof of Existence Theorem 1.3 shows that $\{(\bar{x}_1^{m_v}(t), \ldots, \bar{x}_n^{m_v}(t))\}$ is a uniformly bounded equicontinuous sequence of functions on I. Hence it contains a subsequence which converges uniformly to I to a function $(w_1(t), \ldots, w_n(t))$. By the same arguments used in the proof of Existence Theorem 1.3 it follows that $(w_1(t), \ldots, w_n(t))$ is a solution of (23) on the interval

$$\left(t_0 - \frac{d}{M\sqrt{n}}, t_0 + \frac{d}{M\sqrt{n}}\right)$$

such that

$$w_i(t_0) = x_i^0 \quad (i = 1, \ldots, n)$$

But inequality (33) shows that the solutions $(u_1(t), \ldots, u_n(t))$ and $(w_1(t), \ldots, w_n(t))$ are different functions. This is a contradiction to the hypothesis of uniqueness of solution. $\quad\square$

For explicit computations of solutions or approximations to solutions, it would seem important to measure how fast the sequences converge which occur in the proofs of Existence Theorem 1.1, Existence Theorem 1.2 and the Corollary to Existence Theorem 1.3. Actually questions about rapidity of convergence and choice of a sequence which converges with maximal rapidity are very serious and would lead us deeply into the subject of numerical analysis. As stated in the introduction, we shall (with all due respect for this important subject) not enter a study of these questions.

Next we show how a quick proof of Existence Theorem 1.3 can be given if a technique from functional analysis, the Schauder Fixed Point Theorem (see the Appendix), is used.

Existence Theorem 1.4 *Suppose that f is continuous on D, an open set in (t, x)-space. Let t_0, x^0, a, b, M, A have the same meaning as in the statement of Existence Theorem 1.1. Then there exists a solution $x(t, t_0, x^0)$ of*

$$x' = f(t, x)$$

on $(t_0 - A, t_0 + A)$ *such that* $x(t_0, t_0, x^0) = x^0$.

Proof. Let $J = [t_0 - A, t_0 + A]$ and let $C[J, R^n]$ denote the Banach space of continuous n-vector functions $\phi = (\phi_1, \ldots, \phi_n)$ with domain J such that

$$\|\phi\| = \sup_{t \in J} \sum |\phi_i(t)|$$

Let

$$\mathcal{F} = \{\phi \in C[J, R^n]/\phi(t_0) = x^0, |\phi(t) - x^0| \leq b \; for \; t \in J\}$$

Then it follows easily that $\mathcal{F}$ is a bounded, convex, closed subset of $C[J, R^n]$. Now we define the mapping

$$\mathcal{M} : \mathcal{F} \to C[J, R^n]$$

as follows: if $\phi(t) \in \mathcal{F}$

$$(\mathcal{M}\phi)(t) = x^0 + \int_{t_0}^{t} f[s, \phi(s)]ds$$

From standard theorems of calculus, it follows that if $\phi \in C[J, R^n]$, then $\mathcal{M}\phi \in C[J, R^n]$. Also

$$\mathcal{M}(\mathcal{F}) \subset \mathcal{F} \quad \text{because}$$

$$(\mathcal{M}\phi)(t_0) = x^0$$

and

$$|(\mathcal{M}\phi)(t) - x^0| \leq \int_{t_0}^{t} |f(s, \phi(s))|ds \leq M|t - t_0| \leq MA \leq M\frac{b}{M} = b$$

$\mathcal{M}(\mathcal{F})$ is an equicontinuous set because if $t, \bar{t} \in K$, then

$$| (\mathcal{M}\phi)(t) - (\mathcal{M}\phi)(\bar{t}) | \leq | \int_{\bar{t}}^{t} f(s, \phi(s))ds | \leq M | t - \bar{t} |$$

and M is independent of ϕ. Hence by Ascoli's Theorem, $\mathcal{M}$ is a compact map of $\mathcal{F}$ into itself and by Schauder's Theorem, $\mathcal{M}$ has a fixed point. That is, there is a function $\phi \in \mathcal{F}$ such that

$$\phi(t) = (\mathcal{M}\phi)(t) = x^0 + \int_{t_0}^{t} f[s, \phi(s)]ds$$

By Lemma 1.1, ϕ is the desired solution $x(t, t_0, x^0)$.   □

The brevity of the proof of Existence Theorem 1.4 is misleading because the proof uses the Schauder Fixed Point Theorem, the proof of which is by no means short. Note also that the proof of Existence Theorem 1.4 is non-constructive. That is, we arrive at the conclusion that a solution exists but we obtain no hint about how to compute the solution. If f satisfies a Lipschitz condition, then the sequences constructed in the proofs of Existence Theorems 1.1 and 1.3 converge to solutions. As pointed out earlier, this convergence may be too slow for practical computation of a uniform approximation to a solution. But the solution is approximated in some theoretical sense whereas the proof of Existence Theorem 1.4 yields no approximation at all.

Extension Theorems

In Existence Theorems 1.1, 1.2, 1.3, and 1.4, we have obtained estimates on the size of the domain of the solution. E.g., in Existence Theorem 1.1, the domain of the solution contained the interval $(t_0 - A, t_o + A)$. However none of these estimates put a limitation or bound on the domain of the solution. That is, it might be possible that for a particular equation, the solution were defined for all real t even though the existence theorem only guaranteed the existence of a solution on a finite interval. (We will see later that such a result holds for "well-behaved" linear systems.) Consequently we want now to study further the question of how large the domain of the solution may be.

In the differential equations studied in an elementary course in differential equations, the solutions are often defined for all real t. But it would be quite wrong to conclude that this kind of result generally holds for solutions of differential equations. The following simple example shows a typical complication that may arise. Let x be a 1-vector and consider the equation

$$x' = x^2 \tag{34}$$

By separation of variables, we have

$$\frac{dx}{x^2} = dt$$

Integrating, we obtain

$$-\frac{1}{x} = t + c$$

or

$$x = \frac{-1}{t + c}$$

where c is an arbitrary real constant. Hence the solution $x(t)$ of (34) which satisfies the initial value

$$x(1) = -1$$

is

$$x(t) = -\frac{1}{t}$$

As $t \to 0$, the solution $x(t)$ decreases without bound and solution $x(t)$ is certainly not defined at $t = 0$. Actually no solution of (34) is defined for all real t (except for the solution $x(t) \equiv 0$). This is true in spite of the fact that the function $f(t, x)$ in (34) is just the simple expression x^2 so that $f(t, x)$ is defined on the entire (t, x)-plane and is, indeed, independent of t.

Our next step is to obtain some conditions under which extensions of domains of solutions can be made.

Definition. 1.4 *If $x(t)$ is a solution on (a,b) of $x' = f(t,x)$ and $y(t)$ is a solution on (α, β) of $x' = f(t,x)$ and $(\alpha, \beta) \supset (a,b)$ and $y/(a,b) = x$, then the solution $y(t)$ is an extension of solution $x(t)$.*

Definition. 1.5 *If $x(t)$ is a solution on (a,b) of $x' = f(t,x)$ and if $x(t)$ is such that any extension $y(t)$ of $x(t)$, where $y(t)$ is a solution on (α, β) of $x' = f(t,x)$, has the property that $(\alpha, \beta) = (a,b)$, then $x(t)$ is a maximal solution of $x' = f(t,x)$.*

Extension Theorem 1.1 *Suppose G is a bounded open set in (t, x)-space and f is continuous on $\bar{G}$ and f satisfies a Lipschitz condition in x in G. Suppose $x(t)$, with domain (α, β), where $\beta < \infty$, is a maximal solution of*

$$x' = f(t, x)$$

Let

$$p(t) = \inf_{(t_1, x^1) \in \partial G} \{|t - t_1| + |x(t) - x^1|\}$$

where ∂G denotes the boundary of G. Then $\lim_{t \uparrow \beta} p(t) = 0$.

Remarks.   (1) Extension Theorem 1.1 says roughly that if the solution cannot be extended any further, then it has gone out to the boundary of G.

   (2) The hypothesis that f satisfy a Lipschitz condition in G is to insure uniqueness. (We will point out in the proof where this uniqueness condition is needed.)

Proof of Extension Theorem 1.1. Suppose the conclusion of the theorem is not true. Then there exists $\bar{\varepsilon} > 0$ and a sequence $\{t_m\}$ such that $t_m \uparrow \beta$ and for all m, $p(t_m) > \bar{\varepsilon}$, i.e.,

$$\inf_{(t_1, x^1) \in \partial G} \{|t_m - t_1| + |x(t_m) - x^1|\} > \bar{\varepsilon} \tag{35}$$

Suppose $M > 0$ is such that

$$\max_{\substack{(t,x) \in \bar{G} \\ i=1,\dots,n}} |f_i(t, x)| \leq M$$

Given integer N, then there exists t_m such that

$$|\beta - t_m| < \frac{\bar{\varepsilon}}{N} \tag{36}$$

Let $y(t)$ be a solution of $x' = f(t, x)$ such that

$$y(t_m) = x(t_m)$$

The domain of $y(t)$ contains

$$[t_m - r, t_m + r]$$

where

$$r = \min\{a, \frac{b}{M}\} \qquad (37)$$

and a, b are as described in Existence Theorem 1.1. Let $a = \frac{\bar{\varepsilon}}{N}$. Then we may set b equal to $(\frac{N-1}{N})\bar{\varepsilon} = (1 - \frac{1}{N})\bar{\varepsilon}$ because by (35) the set

$$\left\{ (t,x)/|t_m - t| < \frac{\bar{\varepsilon}}{N}, \; |x(t_m) - x| < \left(1 - \frac{1}{N}\right)\bar{\varepsilon} \right\}$$

is contained in G. But if N is sufficiently large

$$a = \frac{\bar{\varepsilon}}{N} < \left(1 - \frac{1}{N}\right)\frac{\bar{\varepsilon}}{M} = \frac{b}{M} \qquad (38)$$

By (36), (37) and (38), the domain of $y(t)$ contains β. By the Lipschitz condition (See Remark (2) above.) $y(t)$ yields an extension of $x(t)$. Since the domain of the extension contains β, then $x(t)$ is not maximal. This contradicts the hypothesis. $\qquad \square$

Extension Theorem 1.2 *Let G be an open set in (t,x)-space and let f be continuous on $\bar{G}$. Suppose f satisfies a Lipschitz condition in x in G. Let $x(t)$, with domain (α, β) and $\beta < \infty$, be a maximal solution of*

$$x' = f(t,x)$$

Then either $|x(t)|$ becomes unbounded as $t \uparrow \beta$ (i.e., given $\varepsilon > 0$ and $M_1 > 0$, then there exists t such that $0 < \beta - t < \varepsilon$ and $|x(t)| > M_1$) or $\partial G \neq \phi$ and

$$\lim_{t \uparrow \beta} p(t) = 0$$

Proof. Suppose $|x(t)|$ remains bounded as $t \uparrow \beta$. That is, suppose there exists $\varepsilon_0 > 0$ and $c > 0$ such that for all $t \in (\beta - \varepsilon_0, \beta)$,

$$(t, x(t)) \in \{(t,x)/|t| + |x| < c\}$$

Let

$$B_1 = \{(t,x)/|t| + |x| < 2c\}$$

and

$$G_1 = G \cap B_1$$

Then by Extension Theorem 1.1,

$$\liminf_{\substack{(t_1,x^1)\in \partial G_1 \\ t\uparrow\beta}} \{|t - t_1| + |x(t) - x^1|\} = 0 \tag{39}$$

But
$$\partial G_1 \subset \partial G \cup \partial B_1$$

Hence if $(t,x) \in \partial G_1$ and $(t,x) \notin \partial G$, then $(t,x) \in \partial B_1$. But if $(\bar{t},\bar{x}) \in \partial B_1$, and if $t \in (\alpha,\beta)$, then from the definition of B_1, it follows that if $t \in (\beta - \varepsilon_0, \beta)$, then

$$|\bar{t} - t| + |\bar{x} - x(t)| \geq c$$

Hence from (39), we have:

$$\liminf_{\substack{(t_1,x^1)\in \partial G \\ t\uparrow\beta}} \{|t - t_1| + |x(t) - x^1|\} = 0 \qquad \square$$

Extension Theorem 1.3 *Suppose f is continuous on (t,x)-space and that for each point in (t,x)-space, there is a neighborhood N of (t,x) such that in N, f satisfies a Lipschitz condition in x. Suppose also that f is bounded, i.e., suppose there exists $M > 0$ such that for all (t,x),*

$$|f(t,x)| < M$$

Then the domain of each maximal solution of

$$x' = f(t,x)$$

is the t-axis.

Proof. Suppose a maximal solution $x(t)$ has domain (α,β) where $\beta < \infty$. Let $t_0 \in (\alpha,\beta)$. Then by Lemma 1.1,

$$x(t) = x(t_0) + \int_{t_0}^{t} f[s, x(s)]ds$$

and if $\alpha < t_2 < t_1 < \beta$,

$$|x(t_1) - x(t_2)| \le M|t_1 - t_2|$$

Hence the functional values $x(t)$ for t near β and less than β satisfy a Cauchy condition and there exists $\lim_{t \uparrow \beta} x(t)$. Call this limit $\bar{x}$ and let $y(t)$ be a solution of

$$x' = f(t, x)$$

such that $y(\beta) = \bar{x}$. By the Lipschitz condition, $y(t)$ yields an extension of $x(t)$. Since the domain of this extension contains β, then $x(t)$ is not maximal. Contradiction. A similar argument shows that $\alpha > -\infty$ leads to a contradiction. $\quad\square$

(For another version of Extension Theorem 1.3, see Exercise 14.)

Exercises for Chapter 1

1. Justify rigorously the method of separation of variables as applied to the equation: $(*)$ $g(x)\frac{dx}{dt} = f(t)$.

2. If

$$x = \begin{bmatrix} x_1 \\ \vdots \\ x_n \end{bmatrix}$$

   prove that

   $(*)$ $\qquad\qquad \|x\| \le |x|$

   and

   $(**)$ $\qquad\qquad |x| \le \sqrt{n}\|x\|$

3. Let x denote a scalar (i.e., a 1-vector). Prove that the equation

$$\frac{dx}{dt} = \left(\frac{1}{t^2 + 1}\right) e^{-x^2 \sin^2 t}$$

has a unique solution $x(t)$ such that $x(0) = 1$. Show that the domain of $x(t)$ is the real line.

4. Let a, b, c, d be real constants. Show that if $(x(t), x_2(t))$ is a solution of the 2-dimensional system

$$\frac{dx_1}{dt} = ax_1 + bx_2$$

$$\frac{dx_2}{dt} = cx_1 + dx_2$$

such that $x_1(0) = 5$, $x_2(0) = 0$, then for each t in the domain of the solution $(x_1(t), x_2(t))$ it is true that

$$[x_1(t)]^2 + [x_2(t)]^2 > 0$$

(That is, there is no value $\bar{t}$ such that $x(\bar{t}) = y(\bar{t}) = 0$.)

5. Given the system

$$(*) \quad \frac{d^3x}{dt^3} + \left(\frac{1}{1 + t^2}\right)\frac{d^2x}{dt^2} + (\sin t)\frac{dx}{dt} + \frac{t}{x^2 + y^2 + 1} = 0$$

$$\frac{d^2y}{dt^2} + e^{-t}\frac{dy}{dt} + \cos(x + y) = 0$$

show that there exists a unique solution $(x(t), y(t))$ such that $x(0) = 1$, $\frac{dx}{dt}(0) = 0$, $\frac{d^2x}{dt^2}(0) = 5$, $y(0) = 0$, $\frac{dy}{dt}(0) = 3$. Find an interval I on the t-axis with midpoint $t = 0$ such that the domain of the solution contains I.

6. Find the solution of

$$\frac{dy}{dt} = 3t^2 y^2$$

such that $y(1) = 0$.

7. Prove Lemma 1.5 without using the exponential series.

8. Prove Corollary 1.1. (Hint: use Lemma 1.1.)

9. In the Remark after the statement of Existence Theorem 1.3, it is shown that any nonautonomous system of n equations can be transformed into an autonomous system of $(n+1)$ equations. This suggests that the theory of differential equations need only be developed for autonomous systems. (To apply such theory to a nonautonomous system, one would simply transform the nonautonomous system into an autonomous system.) Can you suggest why this possibility might not be feasible?

10. Find $x\left(\frac{3}{4}\right)$ where $x(t)$ is the solution of

$$\frac{dx}{dt} = x^3$$

such that $x(0) = 1$.

11. Find $x\left(\frac{1}{4}\right)$ where $x(t)$ is the solution of

$$\frac{dx}{dt} = x^{1/3}$$

such that $x\left(\frac{1}{2}\right) = 0$.

12. An extremely useful result which will be used in Exercise 13 and used often in Chapter 4, is the following:

Gronwall's Lemma *If u, v are real-valued nonnegative continuous functions with domain $\{t/t \geq t_0\}$ and if there exists a constant $M \geq 0$ such that for all $t \geq t_0$*

$$u(t) \leq M + \int_{t_0}^{t} u(s)v(s)ds \tag{1}$$

then

$$u(t) \leq M \exp \int_{t_0}^{t} v(s)ds \qquad (2)$$

Prove Gronwall's Lemma. (Hint: first assume $M > 0$. Then for all $t \geq t_0$

$$\frac{u(t)v(t)}{M + \int_{t_0}^{t} u(s)v(s)ds} \leq v(t).$$

Integrate both sides of this inequality from t_0 to t.)

13. The Existence Theorem for Equation with a Parameter is a local result. That is, we proved that in a sufficiently small domain

$$\{t/|t - t_0| \leq r\}$$

the solution is a continuous function of parameter μ and initial value x_0. Now suppose that if $\mu = 0$, the equation

$$\frac{dx}{dt} = f(t, x, \mu) \qquad (*)$$

has a unique solution $x(t, \bar{x}^0, 0)$ whose domain contains the interval $[a, b]$. (In all solutions, the initial value will be assumed at $t = t_0$. Hence we will omit the t_0 in the notation for the solution.) Assume that f is continuous and satisfies a Lipschitz condition with Lipschitz constant k on an open set G in (t, x^0, μ)-space where

$$G \supset [a, b] \times \bigcup_{t \in [a,b]} \{x/|x - x(t, \bar{x}_0, 0)| \leq r\} \times [-\mu_0, \mu_0]$$

where r, μ_0 are given positive numbers. Assume also that at each point of G, the function f has a continuous partial derivative $\frac{\partial f}{\partial \mu}$.

Prove that if $|\mu|$ is sufficiently small, equation $(*)$ has a unique solution $x(t, x^0, \mu)$ whose domain contains $[a, b]$ and $x(t, x^0, \mu)$ is continuous in (x^0, μ) uniformly for $t \in [a, b]$.

14. State and prove a version of Extension Theorem 1.3 which does not have a uniqueness hypothesis.

15. Prove: Suppose G is an open set in (t, x)-space and let f be continuous on $\bar{G}$ and be such that for each $(t, x) \in G$ there is a neighborhood N of (t, x) such that in N, f satisfies a Lipschitz condition in x. Let $x(t)$ be a solution with domain (α, β) of the equation

$$(1) \qquad\qquad x' = f(t, x)$$

such that $x(t)$ cannot be extended beyond β and suppose there exists a closed bounded set $A \subset R^n$ such that $[\alpha, \beta] \times A \subset G$ and such that for all $t \in (\alpha, \beta)$

$$x(t) \in A$$

Then $\beta = \infty$.

Examples

Now we describe some differential equations which are used to model biological, physiological, and chemical systems, and indicate how the theory in Chapter 1 is used in the study of these equations. As theory is developed in subsequent chapters, it will be applied to further study of these equations.

1. The Volterra equations for predator-prey systems

For discussions of the derivation of the Volterra equations and other more general cases, see Maynard Smith [1974] and May [1973].

If x denotes the population density of the prey and y the population density of the predator, the Volterra equations describe the rates of change of x and y as follows:

$$\begin{aligned} x' &= ax - Ax^2 - cxy \\ y' &= -dy + exy \end{aligned} \qquad (V)$$

where a, A, c, d and e are positive constants. These equations are sufficiently explicit so that a detailed analysis of their solutions can be made as indicated in Maynard Smith [1974]. Here we will make just a few general observations of a kind that are applicable for more general classes of equations.

Since x, y denote populations and hence are non-negative and remain finite for all $t > t_0$, some fixed value, we need to show that any biologically significant solution of (V) has such properties. That is, we must show that if the solution $(x(t), y(t))$ is such that $(x(t_1), y(t_1))$ is in the first quadrant, then for all $t > t_1$, $(x(t), y(t))$ is defined, is in the first quadrant, and the set

$$\{(x(t), y(t))/t > t_1\}$$

is bounded.

(a) Show that no solution "escapes" the first quadrant, i.e., that if $(x(\bar{t}), y(\bar{t}))$ is in the first quadrant, then there is no value $\tilde{t} > \bar{t}$ such that $(x(\tilde{t}), y(\tilde{t}))$ is not in the first quadrant.

(b) Show that if $\frac{a}{A} < \frac{d}{e}$ and if

$$K = \frac{a}{A} + \delta < \frac{d}{e}$$

where δ is a sufficiently small positive number, then no solution escapes the rectangle R with vertices

$$(0,0), (K,0), (K,B), (0,B)$$

where B is any fixed positive number. Show also that any solution which passes through a point in the first quadrant ultimately enters the rectangle R.

These results show that each solution which passes through a point in the first quadrant ultimately enters and thereafter remains in rectangle R. Hence by Exercise 15, any solution which passes through a point in the first quadrant is defined for all t greater than some t_0.

It should be emphasized that we describe here only some mathematical aspects of the study of the above equations. The more important question of the biological significance of the mathematical results will not be dealt with here at all.

2. The Hodgkin-Huxley equations

One of the most successful mathematical models used in biological sciences is the system of differential equations obtained by Hodgkin and Huxley [1952] in their study of nerve conduction (for which they received a Nobel prize in 1959). For a lucid account of some of the physiological background of their work, see Fitzhugh [1969].

Besides their value as a model in nerve conduction, the Hodgkin-Huxley equations have proved to be very valuable in modeling other physiological systems. They have provided a paradigm, sometimes called Hodgkin-Huxley like equations, for models, among others, of cardiac components (see Cronin [1987]) and brain components (see Traub and Miles [1991]).

The Hodgkin-Huxley equations describe the relationships among the potential difference across the membrane surface of the nerve axon and the current arising from the flow of ions (mostly sodium and potassium) across the membrane and the current caused by the fact that the membrane has a capacitance. For the standard temperature 6.3°C, the H-H equations are the following four equations:

$$\frac{dV}{dt} = \frac{I}{C} - \frac{I_i}{C}$$

or

$$\frac{dV}{dt} = \frac{I}{C} - \frac{1}{C}[\bar{g}_{Na}m^3h(V - V_{Na}) + \bar{g}_K n^4(V - V_K) + \bar{g}_L(V - V_L)]$$

and

$$\frac{dm}{dt} = \alpha_m(V)[1 - m] - \beta_m(V)m = \frac{m_\infty(V) - m}{\tau_m(V)} \qquad \text{(H-H)}$$

$$\frac{dh}{dt} = \alpha_h(V)[1 - h] - \beta_h(V)h = \frac{h_\infty(V) - h}{\tau_h(V)}$$

$$\frac{dn}{dt} = \alpha_n(V)[1 - n] - \beta_n(V)n = \frac{n_\infty(V) - n}{\tau_n(V)}$$

where

t = time

V = potential across membrane

I = total current thru membrane, per unit area

I_i = ionic current thru membrane, per unit area

$\quad = \bar{g}_{Na}m^3 h(V - V_{Na}) + \bar{g}_K n^4(V - V_K) + \bar{g}_L(V - V_L)$

C = capacitance per unit area of membrane (which is a constant)

$\bar{g}_{Na}$ = sodium conductance constant (positive)

$\bar{g}_K$ = potassium conductance constant (positive)

$\bar{g}_L$ = leakage conductance constant (positive)

$$V_{Na} = 115 \text{ mV}$$

$$V_K = -12 \text{ mV}$$

$$V_L = 10.5989 \text{ mV}$$

$$m = \text{sodium activation in H-H model}$$

$$h = \text{sodium inactivation in H-H model}$$

$$n = \text{potassium activation in H-H model}$$

The second expression for $\frac{dm}{dt}$, i.e., $\frac{m_\infty(V)-m}{\tau_m(V)}$, is obtained as follows. Since

$$\frac{dm}{dt} = \alpha_m(V) - [\alpha_m(V) + \beta_m(V)]m = \frac{\frac{\alpha_m(V)}{\alpha_m(V)+\beta_m(V)} - m}{\frac{1}{\alpha_m(V)+\beta_m(V)}}$$

we take

$$m_\infty(V) = \frac{\alpha_m(V)}{\alpha_m(V) + \beta_m(V)}$$

and

$$\tau_m(V) = \frac{1}{\alpha_m(V) + \beta_m(V)}$$

The second expressions for $\frac{dh}{dt}$ and $\frac{dn}{dt}$ are obtained similarly. The variables m, h, n are "phenomenological variables" which describe changes in conductance of sodium and potassium. Their ranges are the interval $[0,1]$. For a discussion of the meaning of m, h, n, see Hodgkin and Huxley [1952] or Cronin [1987].

$$\alpha_m(V) = \frac{.1(25 - V)}{e^{0.1(25-V)} - 1}$$

$$\beta_m(V) = 4e^{-\frac{V}{18}}$$

$$\alpha_h(V) = .07e^{-\frac{V}{20}}$$

$$\beta_h(V) = \frac{1}{e^{.1(30-V)} + 1}$$

$$\alpha_n(V) = \frac{.01(10 - V)}{e^{.1(10-V)} - 1}$$

$$\beta_n(V) = .125e^{-\frac{V}{80}}$$

These equations were obtained from "space-clamped" data, i.e., from experimental data in which V, m, h, n depend on time but not on position along the axon.

Besides giving a mathematical description which summarizes experimental data, the H-H equations make a number of valid predictions. For example, the equations are derived empirically from data in which V is controlled and I and I_i are measured (sometimes called voltage-clamp data). But the equations reproduce or predict data from current-clamp experiments, i.e., experiments in which I is controlled and V is measured. However the II-H equations are obtained by empirical considerations and partly as a consequence of this, their status is quite different from the status of differential equations in mechanics or electrical circuit theory which are derived from first principles such as Newton's laws or the Kirchhoff laws.

The H-H equations present very serious mathematical problems as we will see later. However it is easy to show that the solutions are appropriately bounded.

(a) Show that if I is a bounded continuous function of t there exists a positive number K such that if $(V(t), m(t), h(t), n(t))$ is a solution of (H-H) such that for $t = t_1$

$(V(t), m(t), h(t), n(t)) \in$
$\quad \{(V, m, h, n) / - K \leq V \leq K, m \in [0,1], h \in [0,1], n \in [0,1]\} \quad (*)$

then (∗) holds for all $t \geq t_1$. (That is, the solution is defined for all $t \geq t_1$ and (∗) is true for all $t \geq t_1$.)

It is not difficult to show by combining (H-H) and a few statements from electrical theory that a system of partial differential equations can be derived which describes V, m, h, n as functions of space, i.e., position on the nerve axon, as well as time. (See Cronin [1987], pp. 61-62.) These equations have the following form:

$$\frac{\partial^2 V}{\partial x^2} - a\frac{\partial V}{\partial t} = F(V, m, h, n)$$

$$\frac{\partial m}{\partial t} = G_1(V, m) \tag{P}$$

$$\frac{\partial h}{\partial t} = G_2(V, h)$$

$$\frac{\partial n}{\partial t} = G_3(V, n)$$

where F, G_1, G_2, G_3 are "well-behaved" functions and a is a positive constant. Study of these equations is crucially important because certain of their solutions can be interpreted as descriptions of the impulse which travels along the axon when stimulus is applied. Such solutions, called travelling wave solutions, are of the form

$$V(x - \omega t), \ m(x - \omega t), \ h(x - \omega t), \ n(x - \omega t)$$

where ω is a constant. The search for travelling wave solutions of (P) reduces to the problem of solving a system of ordinary differential equations as the following problem shows.

(b) Let $\xi = x - \omega t$. Show that

$$\frac{\partial^2 V}{\partial x^2}(x - \omega t) = \frac{d^2 V(\xi)}{d\xi^2}$$

and

$$\frac{\partial V}{\partial t} = -\omega\frac{dV}{d\xi}$$

and find the system of ordinary differential equations obtained by making these substitutions in (P).

One of the major successes of the Hodgkin-Huxley theory is that by studying travelling wave solutions, Hodgkin and Huxley were able to make a theoretical estimate of the velocity of nerve conduction which agreed quite well with the experimentally observed velocity.

Since the study of (H-H) and (P) presents serious mathematical difficulties, simpler systems which seem to retain many of the important properties of (H-H) or (P) have been studied instead. One such system which has received considerable study is the FitzHugh-Nagumo equation

$$V_{xx} - V_t = F(V) + R$$
$$R_t = \varepsilon(V - bR)$$

where ε, b are positive constants and $F(V)$ is a function such that $F(0) = 0$ and $F'(0)$ is positive, for example, $F(V)$ can be cubic. For a description of some of the work on the FitzHugh-Nagumo equation as well as other topics in nerve conduction theory, see Scott [1975] and Cronin [1987].

3. The Field-Noyes Model for the Belousov-Zhabotinsky Reaction

The Belousov-Zhabotinsky reaction is a chemical reaction which exhibits tempotal oscillations. A detailed chemical mechanism for the reaction was developed by Field, Köros and Noyes [1972] and this mechanism was described in terms of a differential equation by Field and Noyes [1974]. The mathematical properties of the differential equation have been studied by Hastings and Murray [1975]. The Field-Noyes model can be written as:

$$\frac{dx}{dt} = k_1 Ay - k_2 xy + k_3 Ax - 2k_4 x^2$$
$$\frac{dy}{dt} = -k_1 Ay - k_2 xy + k_5 fz \qquad \text{(F-N)}$$
$$\frac{dz}{dt} = k_3 Ax - k_5 z$$

where $A, f, k_1, k_2, k_3, k_4, k_5$ are positive constants, and x, y, z denote concentrations of certain molecules and ions.

(a) Show that no solution of (F-N) escapes the first octant.

(b) Find a closed bounded set E in the first octant such that no solution of (F-N) escapes it and such that every solution of (F-N) which passes through a point in the first octant ultimately enters and remains in the set E; more precisely, if t_1 is such that $(x(t_1), y(t_1), z(t_1))$ is in the first octant, then there is a number $t_2 > t_1$ such that for all $t \geq t_2$, solution $x(t), y(t), x(t)$ is defined and

$$(x(t), y(t), z(t)) \in E$$

The Belousov-Zhabotinsky reaction also exhibits spatial structure and this can be studied mathematically by using a partial differential equation, i.e., a reaction-diffusion equation. See Tyson [1976], Kopell and Howard [1973].

4. The Goodwin equations for a chemical reaction system

Goodwin [1963, 1965] has introduced and studied a differential equation which describes a chemical reaction system of a type which occurs in the study of cells in biology. The chemical system is assumed to have n constituents and the concentrations $x_1, \ldots, x_n$ of the constituents are related by the equations

$$\frac{dx_1}{dt} = \frac{K}{1 + \alpha x_n^\rho} - b_1 x_1$$
$$\frac{dx_2}{dt} = g_1 x_1 - b_2 x_2 \qquad\qquad \text{(G)}$$
$$\cdots$$
$$\frac{dx_n}{dt} = g_{n-1} x_{n-1} - b_n x_n$$

where ρ is a positive integer, and $\alpha, K, g_1, \ldots, g_{n-1}, b_1, \ldots, b_n$ are positive constants. The term $K/(1 + \alpha x_n^\rho)$ refers to the fact that the reaction dx_1/dt is inhibited by feedback metabolite x_n in a reaction in which the stoichiometric coefficient of x_n is ρ. Since $x_1, \ldots, x_n$ are concentrations we are concerned only with solutions of (G) which are in the first octant.

(a) For $n = 3$, show that no solution of (G) escapes the first octant.

(b) For $n = 3$, determine a closed bounded set E in the first octant such that every solution which passes through a point in the first octant ultimately enters and remains in the set E.

Chapter 2

Linear Systems

Existence Theorems for Linear Systems

Our next step is to study linear differential equations which are a special but very important class of differential equations. That is, we study systems of the form

$$
\begin{aligned}
x_1' &= a_{11}(t)x_1 + \cdots + a_{1n}(t)x_n + u_1(t) \\
x_2' &= a_{21}(t)x_1 + \cdots + a_{2n}(t)x_n + u_2(t) \\
&\cdots \\
x_n' &= a_{n1}(t)x_1 + \cdots + a_{nn}(t)x_n + u_n(t)
\end{aligned}
\tag{1}
$$

in which the right-hand sides of the equations are linear in $x_1, \ldots, x_n$.

The study of such systems is crucially important for the following reasons. First, equations of this form often arise in problems in physics and engineering. Secondly, from the viewpoint of pure mathematics the study is important because an elegant and complete theory is obtained. Finally, as happens in other parts of linear and nonlinear analysis, the theory for linear equations is the basis for much of the study of nonlinear equations.

In a theoretical sense, this chapter contains all the topics on linear equations that are studied in an introductory course in differential equations. However there are a number of useful computational techniques described in the introductory course which we have omitted.

For example, a strategic transformation of variables can change a seemingly impossible equation into one which is easily solved. Consequently this chapter is a supplement to the introductory course. It does not supplant it.

All of the examples discussed at the end of Chapter 1 (Volterra equations, Hodgkin-Huxley equations, etc.) are nonlinear equations. Hence the theory in Chapter 2 is not immediately applicable to the examples. However as we will see later, the linear theory will play an important part in the analysis of the examples, especially in studying the stability of the equilibrium points.

If $A(t)$ denotes the matrix

$$[a_{ij}(t)]$$

and if $u(t)$ denotes the vector

$$\begin{bmatrix} u_1(t) \\ \vdots \\ u_n(t) \end{bmatrix}$$

then the system (1) can be written in terms of vectors and matrices as

$$x' = A(t)x + u(t) \tag{2}$$

Because the properties of matrix $A(t)$ play a very important role in the study of the solutions of (1), we introduce some definitions used in the study of matrix $A(t)$.

Definition.   If $A = [a_{ij}]$ is a constant matrix, the *norm* of A, denoted by $|A|$, is

$$|A| = \sum_{i,j=1}^{n} |a_{ij}|$$

if A and B are constant matrices and x is a constant vector, then it follows easily that:

$$|A + B| \leq |A| + |B|$$
$$|AB| \leq |A||B|$$
$$|Ax| \leq |A||x|$$

Definition. If $A(t) = [a_{ij}(t)]$, the *derivative of* $A(t)$, sometimes denoted by $d/dt\, A(t)$, is

$$\left[\frac{d}{dt} a_{ij}(t)\right]$$

The *integral of* $A(t)$ over $[a, b]$, sometimes denoted by

$$\int_a^b A(t)dt$$

is

$$\left[\int_a^b a_{ij}(t)dt\right]$$

and the *trace of* A, sometimes denoted by tr $a(t)$, is

$$\sum_{i=1}^n a_{ij}(t)$$

Existence Theorem 2.1 for Linear Systems. *If for $i, j = 1, \ldots, n$, each $a_{ij}(t)$ is continuous for all real t and if $u(t)$ is continuous for all real t, then if (t_0, x^0) is an arbitrary point in (t, x)-space, there is a unique solution $x(t, t_0, x^0)$ of equation (1) such that $x(t_0, t_0, x^0) = x^0$ and solution $x(t, t_0, x^0)$ has for its domain the real t-axis.*

Remark. Notice that this theorem shows that these is no extension problem for solutions of linear equations. That is, if the elements of $A(t)$ and $u(t)$ are sufficiently well-behaved (e.g., continuous for all real t), then the solution has for its domain the entire t-axis.

Instead of proving this theorem directly, we prove a somewhat more general theorem which is used less frequently but is of sufficient interest to be presented for its own sake.

Existence Theorem 2.2 for Linear Equations. *Suppose $A(t)$ and $u(t)$ are Riemann integrable functions of t on (a, b), i.e., the Riemann integrals over any interval $[c, d]$ contained in (a, b) of the elements of $A(t)$ and $u(t)$ exist, and suppose there exists a function $k(t)$ with domain (a, b) such that*

1. *$k(t)$ is continuous and bounded on (a, b)*

2. *if* $t \in (a, b)$, *then*

$$|A(t)| \leq k(t)$$

and

$$|u(t)| \leq k(t)$$

Let $t_0 \in (a, b)$ and suppose x^0 is a fixed n-vector. Then equation (1) or (2) has a unique solution $x(t)$ on (a, b) such that $x(t_0) = x^0$ in the following sense: if $t \in (a, b)$, then

$$x(t) = x^0 + \int_{t_0}^t A(s)x(s)ds + \int_{t_0}^t u(s)ds \tag{3}$$

Proof. The underlying idea of the proof is the same as the proof of Existence Theorem 1.1. For $t \in (a, b)$, we define

$$x_0(t) = x^0$$

$$\cdots$$

$$x_{n+1}(t) = x^0 + \int_{t_0}^t A(s)x_n(s)ds + \int_{t_0}^t u(s)ds \qquad (n = 0, 1, 2, \ldots)$$

If $x_n(t)$ is continuous on (a, b), then $A(s)x_n(s)$ is integrable over any interval $[c, d]$ contained in (a, b) and hence $x_{n+1}(t)$ is defined and continuous on (a, b). To show that the $x_n(t)$ converge uniformly, we proceed just as in the proof of the Existence Theorem 1.1. First if $t \in (a, b)$,

$$|x_1(t) - x_0(t)| \leq \int_{t_0}^t \{|A(s)||x^0| + |u(s)|\}ds$$

$$\leq (1 + |x^0|) \int_{t_0}^t k(s)ds$$

Let $K(t) = \int_{t_0}^t k(s)ds$ and assume that for $t \in (a, b)$

$$|x_n(t) - x_{n-1}(t)| \leq (1 + |x^0|)\frac{(K(t))^n}{n!}$$

then

$$|x_{n+1}(t) - x_n(t)| \leq \int_{t_0}^{t} |A(s)x_n(s) - A(s)x_{n-1}(s)| ds$$

$$\leq (1 + |x^0|) \int_{t_0}^{t} k(s) \frac{(K(s))^n}{n!} ds$$

Since

$$\frac{d}{dt} K(t) = k(t)$$

and $K(t_0) = 0$, then

$$|x_{n+1}|(t) - x_n(t) \leq (1 + |x^0|) \frac{(K(t))^{n+1}}{(n+1)!}$$

Thus $\{x_n(t)\}$ converges uniformly on any closed interval $[c, d]$ in (a, b) to a continuous function $x(t)$. To complete the proof of existence of the solution, it is sufficient to show that

$$\lim_{n} \int_{t_0}^{t} A(s)x_n(s) ds = \int_{t_0}^{t} A(s)x(s) ds$$

Since $x(t)$ is continuous, $\int_{t_0}^{t} A(s)x(s) ds$ exists and

$$\left| \int_{t_0}^{t} A(s)[x_n(s) - x(s)] ds \right| \leq \int_{t_0}^{t} |A(s)| \, |x_n(s) - x(s)| ds \leq \varepsilon M$$

where M is a bound for $k(t)$ on (a, b).

The proof that $x(t)$ is a unique solution in any closed interval $[c, d]$ in (a, b) goes through just as for the proof of Lemma 1.5 in the proof of Existence Theorem 1.1. □

The proof of Existence Theorem 2.1 for Linear Systems is obtained from Existence Theorem 2.2 for Linear Systems as follows. If the elements of $A(t)$ and $u(t)$ are continuous then since the solution $x(t)$ is continuous, equation (3) may be differentiated with respect to t and we obtain:

$$\frac{d}{dt} x(t) = A(t)x(t) + u(t)$$

Homogeneous Linear Equations:
General Theory

Definition. If $u(t)$ in equation (2) is identically zero, i.e., if (2) has the form

$$x' = A(t)x \tag{4}$$

then the equation is said to be a *homogeneous* linear equation.

Throughout this discussion we will study the n-dimensional equation (4) and it will be assumed that the elements of matrix $A(t)$ are continuous on an interval (a,b).

Theorem 2.1 *If $x(t)$ is a solution of (4) on (a,b) and if there exists $t_0 \in (a,b)$ such that $x(t_0) = 0$, then $x(t) = 0$ for all $t \in (a,b)$.*

Proof. Let $y(t) = 0$ for $t \in (a,b)$. It is clear that $y(t)$ is a solution of (4) on (a,b). But if $t = t_0$, $x(t) = y(t)$. Hence by the uniqueness condition in Existence Theorem 2.1 for Linear Systems, $x(t) = y(t)$ for all $t \in (a,b)$.   □

Definition. Let $f_1(t), \ldots, f_q(t)$ be n-vector functions on (a,b). Then $f_1(t), \ldots, f_q(t)$ are *linearly dependent* if there exist constants $c_1, \ldots, c_q$ (not all zero) such that for all $t \in (a,b)$,

$$c_1 f_1(t) + \cdots + c_q f_q(t) = 0$$

If $f_1(t), \ldots, f_q(t)$ are not linearly dependent, they are *linearly independent*.

Theorem 2.2 *The collection of solutions of (4) is an n-dimensional linear space.*

Proof. The collection of solution is clearly a linear space. Let $t_0 \in (a,b)$ and let

$$x^0_{(1)} = (1, 0, \ldots, 0)$$
$$x^0_{(2)} = (0, 1, 0, \ldots, 0)$$
$$\cdots$$
$$x^0_{(n)} = (0, \ldots, 0, 1)$$

By the existence theorem, there are solutions $x^{(i)}(t)$ of (4) such that

$$x^{(i)}(t_0) = x^0_{(i)}$$

But these solutions are linearly independent because suppose there are constants $c_1, \ldots, c_n$ such that for $t \in (a, b)$

$$c_1 x^{(1)}(t) + c_2 x^{(2)}(t) + \cdots + c_n x^{(n)}(t) = 0$$

Then

$$c_1 x^{(1)}(t_0) + c_2 x^{(2)}(t_0) + \cdots + c_n x^{(n)}(t_0) = 0$$

or

$$(c_1, c_2, \ldots, c_n) = 0$$

or

$$c_1 = c_2 = \cdots = c_n = 0$$

Thus the collection of solutions is a linear space of dimension greater than or equal to n.

Now suppose $y^{(1)}(t), y^{(2)}(t), \ldots, y^{(n)}(t), y^{(n+1)}(t)$ are linearly independent solutions of (4). Let $t_0 \in (a, b)$. Then by Theorem 2.1, $y^{(1)}(t_0), y^{(2)}(t_0), \ldots, y^{(n)}(t_0), y^{(n+1)}(t_0)$ are linearly independent. That is, we have $(n + 1)$ linearly independent n-vectors. From the theory of linear algebra, this is impossible. $\square$

Definition. A set of n linearly independent solutions of (4) is a *fundamental system* of (4).

Definition. An $n \times n$ matrix whose n columns are n linearly independent solutions of (4) is a *fundamental matrix* of (4).

Theorem 2.3 *Let X be an $n \times n$ matrix whose columns are solutions of (4). A necessary and sufficient condition that X be a fundamental matrix of (4) is that there exist $t_0 \in (a, b)$ such that $\det X(t) \neq 0$ at $t = t_0$.*

Proof. If $\det X(t) \neq 0$ at $t = t_0$, then the columns of $X(t_0)$ are linearly independent. Then by Theorem 2.1, the columns of $X(t)$ are linearly independent for each $t \in (a, b)$.

If the columns of $X(t)$ are linearly independent for each $t \in (a, b)$, then $\det X(t) \neq 0$ for each $t \in (a, b)$. This completes the proof of Theorem 2.3. $\square$

Theorem 2.4 *If X is a fundamental matrix of (4) and C is a constant nonsingular matrix, then XC is a fundamental matrix. If X_1 is a second fundamental matrix, there exists a constant nonsingular matrix C_1 such that $X_1 = XC_1$.*

Proof. Since

$$X' = A(t)X$$

then

$$X'C = A(t)\{XC\}$$

or

$$(XC)' = A(t)\{XC\}$$

Also

$$\det XC = (\det X)(\det C) \neq 0$$

Thus XC is a fundamental matrix. For $t \in (a, b)$, let

$$(X(t))^{-1}(X_1(t)) = Y(t)$$

Then

$$X_1(t) = X(t)Y(t)$$

and

$$X_1' = XY' + X'Y$$

or

$$AX_1 = XY' + AXY = XY' + AX_1$$

Therefore

$$XY' = 0$$

Since $\det X(t) \neq 0$ for each $t \in (a, b)$, then for each $t \in (a, b)$

$$Y'(t) = 0$$

Therefore $Y(t)$ is a constant matrix. Since

$$\det Y(t) = \det\{[X(t)]^{-1}\} \det\{X_1(t)\} \neq 0$$

then $Y(t)$ is nonsingular.   □

Theorem 2.5 *If $Y(t)$ is a matrix such that*

$$Y'(t) = A(t)Y(t) \tag{5}$$

then

$$\frac{d}{dt}[\det Y(t)] = [tr\, A(t)][\det Y(t)] \tag{6}$$

Also if $t, t_0 \in (a, b)$, then

$$\det Y(t) = [\det Y(t_0)] \exp\left\{ \int_{t_0}^{t} tr\, A(s)ds \right\}$$

Proof. Let $Y(t) = (y_{ij}(t))$ and $A(t) = (a_{ij}(t))$. Then

$[\det Y(t)]'$

$$= \begin{vmatrix} y'_{11} & y'_{12} & \cdots & y'_{1n} \\ & \vdots & & \\ y_{n1} & y_{n2} & \cdots & y_{nn} \end{vmatrix} + \cdots + \begin{vmatrix} y_{11} & y_{12} & \cdots & y_{1n} \\ & \vdots & & \\ y'_{n1} & y'_{n2} & \cdots & y'_{nn} \end{vmatrix} \tag{7}$$

But by (5)

$$y'_{ij}(t) = \sum_{k=1}^{n} a_{ik}(t)y_{kj}(t)$$

Then (7) becomes

$$[\det Y(t)]' = \begin{vmatrix} \sum a_{1k}(t)y_{k1}(t) & \cdots & \sum a_{1k}(t)y_{kn}(t) \\ y_{21} & & y_{2n} \\ \vdots & & \vdots \\ y_{n1} & & y_{nn} \end{vmatrix} + \cdots$$

In the first determinant on the right, subtract from the first row the expression [a_{12} times second row $+a_{13}$ times third row $+ \cdots +$ a_{1n} times nth row]. Carrying out similar operations on the other determinants on the right, we obtain

$$[\det Y(t)]' = \begin{vmatrix} a_{11}y_{11} & \cdots & a_{11}y_{1n} \\ y_{21} & & y_{2n} \\ \vdots & & \vdots \\ y_{n1} & & y_{nn} \end{vmatrix} + \cdots$$

$$= a_{11}\det[Y(t)] + \cdots + a_{nn}\det[Y(t)]$$

$$= [tr\, A(t)][\det Y(t)]$$

A straightforward computation shows that

$$[\det Y(t_0)] \exp \left\{ \int_{t_0}^{t} tr A(s)ds \right\}$$

is a solution of the scalar differential equation

$$x' = (tr A(t))x \tag{8}$$

But (6) shows that $\det Y(t)$ is a solution of (8). Hence by the uniqueness of the solution we must have

$$\det Y(t) = [\det Y(t_0)] \exp \left\{ \int_{t_0}^{t} tr A(s)ds \right\} \qquad \square$$

Corollary. *If the columns of X are solutions of (4), then X is a fundamental matrix if and only if $\det X(t) \neq 0$ for all $t \in (a, b)$.*

Now we obtain a more explicit representation for the solutions of some linear homogeneous equations by introducing the exponential of a matrix. If $B(t)$ is an $n \times n$ matrix such that the elements are functions of t consider the sums

$$\sum_{s=0}^{m} \frac{B^s}{s!}$$

where $B^0 = I$, the $n \times n$ identity matrix. For each t, these sums satisfy a Cauchy Condition in the matrix norm because

$$\left| \sum_{s=0}^{p+q} \frac{B^s}{s!} - \sum_{s=0}^{p} \frac{B^s}{s!} \right| = \left| \sum_{s=p+1}^{p+q} \frac{B^s}{s!} \right| \leq \sum_{s=p+1}^{p+q} \frac{|B|^s}{s!}$$

But if p is sufficiently large, this last term is $< \varepsilon$ because it is part of the series expansion for $e^{|B|}$. From the definition of the matrix norm, it follows that the elements in the i, jth position in the sums $\sum_{s=0}^{m} \frac{B^s}{s!}$ ($m = 1, 2, 3, \dots$) satisfy a Cauchy condition and therefore converge to a function $c_{ij}(t)$ and the matrices $\sum_{s=0}^{m} \frac{B^s}{s!}$ converge in the matrix norm to the matrix $[c_{ij}(t)]$.

Definition. The matrix $[c_{ij}(t)]$ is called the *exponential* of $B(t)$ and is denoted by $e^{B(t)}$ or $\exp B(t)$.

Lemma 2.1 *If* $BD = DB$, *then* $e^{B+D} = e^B e^D$.

Proof. If b, d are real numbers, then

$$e^{b+d} = e^b e^d$$

or

$$1 + (a+b) + \frac{(a+b)^2}{2!} + \cdots = \left(1 + a + \frac{a^2}{2!} + \cdots\right)\left(1 + b + \frac{b^2}{2!} + \cdots\right) \quad (9)$$

Since $BD = DB$ then (9) is valid with b and d replaced by B and D respectively. $\square$

Theorem 2.6 *Given*

$$x' = A(t)x \quad (10)$$

let $B(t) = \int_{t_0}^t A(s)ds$ *for* t_0, $t \in (a, b)$ *and assume that for each* $t \in (a, b)$,

$$A(t)B(t) = B(t)A(t)$$

Then the solution $x(t)$ *of (10) such that* $x(t_0) = x^0$ *is*

$$x(t) = (\exp(B))x^0$$

Proof. We have already used the fact that the theorem for derivative of a product holds for the derivative of a product of matrices. That is, if $C = (c_{ij}(t))$, $D = (d_{ij}(t))$ and $c_{ij}(t)$, $d_{ij}(t)$ are differentiable functions $(i, j = 1, \ldots, n)$ then

$$\frac{d}{dt}(CD) = C\frac{dD}{dt} + \frac{dC}{dt}D \quad (11)$$

But we need to emphasize that since the multiplication of matrices is not commutative it is essential in using (11) to preserve the order in which the factors occur. From the definition of B, it follows that

$$\frac{d}{dt}B(t) = A(t)$$

Now assume that

$$\frac{d}{dt}[B(t)]^{m-1} = (m-1)A(t)[B(t)]^{m-2}$$

Then using (11) and the hypothesis that A and B commute, we have:

$$\frac{d}{dt}[B(t)]^m = \frac{d}{dt}\{(B(t))^{m-1}(B(t))\}$$
$$= (B(t))^{m-1}A(t) + (m-1)A(t)(B(t))^{m-2}B(t)$$
$$= mA(t)(B(t))^{m-1} \tag{12}$$

Hence by induction, equation (12) holds for all positive integers m. By Existence Theorem 2.1 for Linear Equations, the desired solution $x(t)$ is the limit of the sequence

$$x_1(t) = x^0 + \int_{t_0}^{t} A(s)x^0 ds = (I + B)x^0$$

$$\cdots$$

$$x_m(t) = x^0 + \int_{t_0}^{t} A(s)x_{m-1}(s)ds$$

Then

$$x'_m(t) = A(t)x_{m-1}(t) \tag{13}$$

Assume that

$$x_{m-1}(t) = \left(I + B + \frac{B^2}{2!} + \cdots + \frac{B^{m-1}}{(m-1)!}\right)x^0$$

Then by (13)

$$x'_m(t) = Ax^0 + ABx^0 + \frac{AB^2}{2!}x^0 + \cdots + \frac{AB^{m-1}}{(m-1)!}x^0$$

and by (12)

$$x_m(t) = K + \left(B + \frac{B^2}{2!} + \cdots + \frac{B^m}{m!}\right)x^0 \tag{14}$$

where K is an arbitrary constant vector. Since $B(t_0) = 0$, then if $t = t_0$, (14) yields

$$x_m(t_0) = K$$

But $x_m(t_0) = x^0$. Therefore (14) becomes

$$x_m(t) = x^0 + \left(B + \frac{b^2}{2!} + \cdots + \frac{B^m}{m!}\right)x^0$$

and we have

$$\lim_{m \to \infty} x_m(t) = e^B x^0$$

This completes that proof of Theorem 2.6.   □

The condition that A and B commute is very strong, but there is one important class of equations for which this condition is satisfied, i.e., the case in which A is a constant matrix. Our next objective is to obtain an explicit form for a fundamental matrix for (4) in the case where A is a constant.

Homogeneous Linear Equations With Constant Coefficients

Theorem 2.7 *If A is a constant matrix then e^{At} is a fundamental matrix for*

$$x' = Ax \tag{15}$$

Proof. First, e^B is a fundamental matrix by Theorems 2.3 and 2.6. But by Lemma 2.1,

$$e^B = e^{A(t-t_0)} = e^{At} e^{-At_0}$$

Since e^{At_0} is nonsingular (its inverse is e^{At_0}) then by Theorem 2.4, e^{At} is a fundamental matrix.   □

To study the form of e^{At}, we will use a canonical form of matrix A. So we first summarize briefly the results we shall need from the theory of canonical forms.

Definition. If A is an $n \times n$ constant matrix and I is the identity $n \times n$ matrix, the roots of the equation $\det(A - \lambda I) = 0$ are the *characteristic roots* (or *eigenvalues* or *proper values*) of A. If λ is a root of multiplicity m of the equation $\det(A - \lambda I) = 0$, then λ is an *eigenvalue of A of algebraic multiplicity m.* The equation

$$\det(A - \lambda I) = 0$$

is the *characteristic equation of A.* If x is a nonzero vector such that $(A - \lambda I)x = 0$ or $Ax = \lambda x$, then x is an *eigenvector associated with eigenvalue λ.*

Definition. A real [complex] $n \times n$ matrix C is *similar over the real [complex] numbers* to a real [complex] $n \times n$ matrix D if there exists a nonsingular real [complex] matrix P such that $D = PCP^{-1}$.

(It is easy to verify that similarity is an equivalence relation.)

Lemma 2.2 *If C is similar to D, then the set of characteristic roots of C is the set of characteristic roots of D.*

Proof. If λ_1 is a characteristic root of C, then $\det(C - \lambda_1 I) = 0$. But

$$
\begin{aligned}
\det(D - \lambda_1 I) &= \det(PCP^{-1} - \lambda_1 PP^{-1}) \\
&= \det(P(C - \lambda_1 I)P^{-1}) \\
&= \det P(\det(C - \lambda_1 I)) \det P^{-1} \\
&= 0 \qquad \square
\end{aligned}
$$

Jordan Canonical Form Theorem. *If A is a complex $n \times n$ matrix, then there is a nonsingular complex matrix P such that*

$$
P^{-1}AP = J
$$

where the matrix J has the following form:

$$
J = \begin{bmatrix} J_1 & & & \\ & J_2 & & \\ & & \ddots & \\ & & & J_s \end{bmatrix}
$$

where all entries not written in are zero and J_i, $i = 1, \ldots, s$, has the form

$$
J_i = \begin{bmatrix} \lambda_i & 1 & & & \\ & \lambda_i & 1 & & \\ & & \ddots & 1 \\ & & & \ddots & \\ & & & & \lambda_i \end{bmatrix}
$$

where λ_i is an eigenvalue of A. Each eigenvalue of A appears in at least one J_i. Matrix J_i may be a 1×1 matrix. (In the case that each J_i is a 1×1 matrix, matrix J is a diagonal matrix.) The eigenvalues λ_i, $i = 1, \ldots, s$, are, in general, not distinct. Except for the order in which the matrices J_i $(i = 1, \ldots, s)$ appear on the diagonal, the matrix J is unique. (The matrix (J) is called the Jordan Canonical Form *or the* rational canonical form *of matrix A.)*

Proof. The proof of this theorem is quite lengthy and can be found in any standard text on linear algebra. See, for example, Halmos [1958]. We describe the basic idea of the proof for the special case of an $n \times n$ matrix A which has just one eigenvalue λ. (Then λ has algebraic multiplicity n.) Let $v_1, \ldots, v_k$ be a basis for the linear space of solutions of the equation

$$(A - \lambda I)v = 0$$

That is, $v_1, \ldots, v_k$ is a basis for the space of eigenvectors associated with eigenvalue λ. Let $v_i^{(1)}$ denote a solution of the equation

$$(A - \lambda I)v = v_i$$

if it exists; and if j is an integer such that $j \geq 2$, let $v_i^{(j)}$ denote a solution of the equation

$$(A - \lambda I)v = v_i^{(j-1)}$$

if it exists. (These are called *generalized eigenvectors.*) It can be proved that there are n such vectors $v_1, \ldots, v_k$, $v_1^{(1)}, \ldots$ which are linearly independent. Suppose, for example, that the n vectors are

$$v_1, \ldots, v_k, v_1^{(1)}, v_\ell^{(1)}, v_k^{(1)}, v_1^{(2)} \tag{16}$$

where ℓ is an integer such that $1 < \ell < k$. Let P be the $n \times n$ matrix whose columns are the vectors

$$v_1, v_1^{(1)}, v_1^{(2)}, v_2, \ldots, v_\ell, v_\ell^{(1)}, v_{\ell+1}, \ldots, v_k, v_k^{(1)}$$

Since the vectors are linearly independent, P is nonsingular. Now

$$AP = \lambda v_1, \lambda v_1^{(1)} + v_1, \lambda v_1^{(2)} + v_1^{(1)}, \ldots, \lambda v_\ell, \lambda v_\ell^{(1)} + v_\ell, \ldots, \lambda v_k, \lambda v_k^{(1)} + v_k$$

and therefore

$$P^{-1}AP = P^{-1} \left[\lambda v_1, \lambda v_1^{(1)}, \lambda v_1^{(2)}, \ldots, \lambda v_\ell, \lambda v_\ell^{(1)}, \ldots, \lambda v_k, \lambda v_k^{(1)} \right]$$
$$+ P^{-1} \left[0, v_1, v_1^{(1)}, 0, \ldots, 0, v_\ell, \ldots, 0, v_k \right]$$
$$= P^{-1}\lambda P + P^{-1} \left[0, v_1, v_1^{(1)}, 0, \ldots, 0, v_\ell, \ldots, 0, v_k \right]$$

$$= \lambda I + \begin{bmatrix} 0 & 1 & 0 & \cdots & 0 & \cdots & 0 \\ & & 1 & & & & \\ & & & \ddots & & & \\ & & & \cdots & 1 & \cdots & \\ & & & & & \ddots & \\ & & & & & & 1 \\ 0 & 0 & 0 & \cdots & 0 & & 0 \end{bmatrix}$$

$$\underset{\uparrow\ \rule{1.5cm}{0.4pt}\ (\ell+3)\text{th column}}{}$$

$$= \begin{bmatrix} \lambda & 1 & 0 & & & & & \\ & \lambda & 1 & & & & & \\ & & \lambda & & & & & \\ & & & \ddots & & & & \\ & & & & \lambda & 1 & & \\ & & & & & \lambda & & \\ & & & & & & \ddots & \\ & & & & & & \lambda & 1 \\ & & & & & & & \lambda \end{bmatrix}$$

$$\underset{\uparrow\ \rule{1.5cm}{0.4pt}\ (\ell+3)\text{th column}}{}$$

If the matrix A has more than one eigenvalue, the matrix P is constructed by finding sets of vectors (16) for each eigenvalue. It can be proved that the collection of all such sets of vectors consists of n linearly independent vectors.

The Jordan Canonical Form Theorem says that A is similar over the complex numbers to the matrix J. In most of our work we will be interested in matrices A which are real. Since the eigenvalues of A need not be real, the matrix J may have complex entries, and a real canonical form is sometimes preferable.

Real Canonical Form Theorem.   *If A is a real $n \times n$ matrix,* *then there is a real nonsingular matrix P such that $P^{-1}AP = J$ where matrix J is a real matrix which has the form*

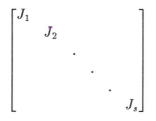

where all entries not written in are zero and J_j, $j = 1, \ldots, s$, is associated with eigenvalue λ_j and has one of the two following forms. If λ_j is real,

$$J_j = \begin{bmatrix} \lambda_j & 1 & & & \\ & \lambda_j & \cdot & & \\ & & \cdot & \cdot & \\ & & & \cdot & 1 \\ & & & & \lambda_j \end{bmatrix}$$

If λ_j is a complex eigenvalue,

$$\lambda_j = \alpha_j + i\beta_j$$

where α_j, β_j are real and $\beta_j > 0$, then

$$J_j = \begin{bmatrix} \alpha_j & \beta_j & 1 & 0 & & & & \\ -\beta_j & \alpha_j & 0 & 1 & & & & \\ 0 & 0 & \cdot & & & & & \\ 0 & 0 & & \cdot & & & & \\ & & & & \cdot & & & \\ & & & & & \alpha_j & \beta_j & 1 & 0 \\ & & & & & -\beta_j & \alpha_j & 0 & 1 \\ & & & & & 0 & 0 & \alpha_j & \beta_j \\ & & & & & 0 & 0 & -\beta_j & \alpha_j \end{bmatrix}$$

If λ_j is real, matrix J_j may be a 1×1 matrix. If λ_j is complex, then J_j must be a $(2m \times 2m)$ matrix where $m \geq 1$. The $\lambda_1, \ldots, \lambda_s$ are not *necessarily distinct. There is associated with each eigenvalue of A at least one J_j. (Note: If $\lambda_j = \alpha_j + i\beta_j$, where $\beta_j > 0$, is an eigenvalue then (since A is real) $\alpha_j - i\beta_j$ is also an eigenvalue. We say that $\alpha_j + i\beta_j$ and $\alpha_j - i\beta_j$ are both associated with J_j.)*

Proof. As in the Jordan canonical form, we simply describe the basic idea of the proof. First suppose that A is a $(2m \times 2m)$ matrix which has $\alpha + i\beta$ as an eigenvalue of algebraic multiplicity m. Since A is real, then $\alpha - i\beta$ is also an eigenvalue of algebraic multiplicity m. It can be proved that the eigenvectors associated with $\alpha + i\beta$ have the form $u_j + iv_j$, $j = 1, \ldots, k$ where $u_1, \ldots, u_k, v_1, \ldots, v_k$ are real and linearly independent. Moreover $u_j - iv_j$ is an eigenvector associated with $\alpha - i\beta$. Let $u_j^{(1)} + iv_j^{(1)}$, where $u_j^{(1)}$ and $v_j^{(1)}$ are real, denote a solution of the equation

$$(A - \lambda I)w = u_j + iv_j$$

(if it exists) and if q is an integer such that $q \geq 2$ let $u_j^{(q)} + iv_j^{(q)}$ denote a solution of the equation

$$(A - \lambda I)w = u_j^{(q-1)} + iv_j^{(q-1)}$$

(if it exists). It can be proved that there are exactly $2m$ vectors $u_1, \ldots, u_k, v_1, \ldots, v_k, u_1^{(1)}, \ldots, v_1^{(1)}, \ldots$ which are linearly independent. Suppose the $2m$ vectors are:

$$u_1, u_1^{(1)}, u_1^{(2)}, u_2, u_2^{(1)}, u_3, \ldots, u_k, v_1, v_1^{(1)}, v_1^{(2)}, v_2, v_2^{(1)}, v_3, \ldots, v_k$$

Let P be the matrix

$$u_1, v_1, u_1^{(1)}, v_1^{(1)}, u_1^{(2)}, v_1^{(2)}, u_2, v_2, u_2^{(1)}, v_2^{(1)}, u_3, v_3, \ldots, u_k, v_k$$

and observe that

$$[A - (\alpha + i\beta)I][u_1^{(2)} + iv_1^{(2)}] = u_1^{(1)} + iv_1^{(1)}$$

or

$$Au_1^{(2)} = \alpha u_1^{(2)} - \beta v_1^{(2)} + u_1^{(1)}$$
$$Av_1^{(2)} = \beta u_1^{(2)} + \alpha v_1^{(2)} + v_1^{(1)}$$

Matrix $A - (\alpha + i\beta)I$ acts similarly on the other vectors. Hence we obtain:

$$
\begin{aligned}
AP = [&\alpha u_1 - \beta v_1, \beta u_1 + \alpha v_1, \alpha u_1^{(1)} - \beta v_1^{(1)} + u_1, \beta u_1^{(1)} + \alpha v_1^{(1)} + v_1, \\
&\alpha u_1^{(2)} - \beta v_1^{(2)} + u_1^{(1)}, \beta u_1^{(2)} + \alpha v_1^{(2)} + v_1^{(1)}, \\
&\alpha u_2 - \beta v_2, \beta u_2 + \alpha v_2, \\
&\alpha u_2^{(1)} - \beta v_2^{(1)} + u_2, \beta u_2^{(1)} + \alpha v_2^{(1)} + v_2, \\
&\alpha u_3 - \beta v_3, \beta u_3 + \alpha v_3, \ldots, \alpha u_k - \beta v_k, \beta u_k + \alpha v_k]
\end{aligned}
$$

Hence

$$
\begin{aligned}
P^{-1}AP = \ &P^{-1}[\alpha u_1, \alpha v_1, \alpha u_1^{(1)}, \alpha v_1^{(1)}, \alpha u_1^{(2)}, \alpha v_1^{(2)}, \alpha u_2, \alpha v_2 \\
&\quad \alpha u_2^{(1)}, \alpha v_2^{(1)}, \alpha u_3, \alpha v_3, \ldots, \alpha u_k, \alpha v_k] \\
&+ P^{-1}[-\beta v_1, \beta u_1, -\beta v_1^{(1)}, \beta u_1^{(1)}, -\beta v_1^{(2)}, \beta u_1^{(2)}, -\beta v_2, \beta u_2, \\
&\quad -\beta v_2^{(1)}, \beta u_2^{(1)}, -\beta v_3, \beta u_3, \ldots, -\beta v_k, \beta u_k] \\
&+ P^{-1}[0, 0, u_1, v_1, u_1^{(1)}, v_1^{(1)}, 0, 0, u_2, v_2, 0, 0, \ldots, 0, 0]
\end{aligned}
$$

From the definition of P and P^{-1}, it follows that

$$
P^{-1}AP = \alpha I +
\begin{bmatrix}
0 & \beta & 0 & 0 & \cdots & & & \\
-\beta & 0 & 0 & 0 & \cdots & & & \\
0 & 0 & 0 & \beta & \cdots & & & \\
& & -\beta & 0 & \cdots & & & \\
& & & & & & 0 & \beta \\
0 & 0 & 0 & 0 & & & -\beta & 0
\end{bmatrix}
$$

$$
+
\begin{bmatrix}
0 & 0 & 1 & 0 & 0 & 0 & \cdots & 0 \\
0 & 0 & 0 & 1 & 0 & 0 & \cdots & 0 \\
& & & 0 & 1 & 0 & \cdots & 0 \\
& & & 0 & 0 & 1 & \cdots & 0 \\
& & & & & & \ddots & \vdots \\
& & & & & & \ddots & 0 \\
& & & & & & \ddots & 0 \\
0 & 0 & & & & & & 0
\end{bmatrix}
$$

For more general kinds of matrices A, extensions of the above idea can be made as in the proof of the Jordan Canonical Form Theorem.

Now let us see what the fundamental matrix e^{tA} looks like. First we apply the Jordan Canonical Form Theorem. Since $A = PJP^{-1}$, then

$$e^{tA} = e^{tPJP^{-1}}$$

$$= I + tPJP^{-1} + \frac{t^2}{2!}(PJP^{-1}) + \cdots + \frac{t^n}{n!}(PJP^{-1})^n + \cdots$$

$$= PP^{-1} + PtJP^{-1} + P\frac{t^2}{2!}JP^{-1} + \cdots + P\frac{t^n}{n!}J^nP^{-1} + \cdots$$

$$= Pe^{tJ}P^{-1}$$

Since $e^{tA} = Pe^{tJ}P^{-1}$ is a fundamental matrix, then by Theorem 2.4 $Pe^{tJ}P^{-1}(P) = Pe^{tJ}$ is also a fundamental matrix. To study the form of Pe^{tJ}, let us first see what e^{tJ} looks like. It follows from a simple observation about matrix multipliction that

$$e^{tJ} = \exp \begin{bmatrix} tJ_1 & & \\ & tJ_2 & \\ & & tJ_s \end{bmatrix} = \begin{bmatrix} e^{tJ_1} & & \\ & e^{tJ_2} & \\ & & e^{tJ_s} \end{bmatrix}$$

where, as usual, entries which are not written are zeros. If J_i is a 1×1 matrix, then e^{tJ_i} is just the scalar $e^{t\lambda_i}$. If J_i is an $r \times r$ matrix where $r > 1$, then

$$J_i = \lambda_i I + D$$

where I is the $r \times r$ identity matrix and

$$D = \begin{bmatrix} 0 & 1 & 0 & \cdots & 0 \\ 0 & 0 & 1 & \cdots & 0 \\ & & \cdots & & \\ 0 & 0 & 0 & \cdots & 1 \\ 0 & 0 & 0 & \cdots & 0 \end{bmatrix}$$

Since $ID = DI$, then by Lemma 2.1,

$$e^{tJ_i} = e^{t(\lambda_i I + D)} = e^{t\lambda_i I} e^{tD}$$

but

$$e^{tD} = \begin{bmatrix} 1 & 0 & \cdots & 0 \\ 0 & 1 & \cdots & 0 \\ & \cdots & & \\ 0 & & & 1 \end{bmatrix} + \begin{bmatrix} 0 & t & 0 & \cdots & 0 \\ 0 & 0 & t & \cdots & 0 \\ & \cdots & & & \\ 0 & 0 & 0 & \cdots & t \\ 0 & 0 & 0 & \cdots & 0 \end{bmatrix}$$

$$+ \frac{1}{2!} \begin{bmatrix} 0 & t & 0 & \cdots & 0 \\ 0 & 0 & t & \cdots & 0 \\ & \cdots & & & \\ 0 & 0 & 0 & \cdots & t \\ 0 & 0 & 0 & \cdots & 0 \end{bmatrix}^2 \qquad (17)$$

$$+ \cdots + \frac{1}{n!} \begin{bmatrix} 0 & t & 0 & \cdots & 0 \\ 0 & 0 & t & \cdots & 0 \\ & \cdots & & & \\ 0 & 0 & 0 & \cdots & t \\ 0 & 0 & 0 & \cdots & 0 \end{bmatrix}^n + \cdots$$

and a simple computation shows that if J_i is an $r \times r$ matrix, then if $h \geq r$, $D^h = 0$. Thus (17) is an easily computed finite sum and we have

$$e^{tJ_i} = \begin{bmatrix} e^{t\lambda_i} & & & \\ & \cdot & & \\ & & \cdot & \\ & & & \cdot \\ & & & & e^{t\lambda_i} \end{bmatrix} \begin{bmatrix} 1 & t & \frac{t^2}{2!} & \cdots & \frac{t^{r-1}}{(r-1)!} \\ & 1 & t & \frac{t^2}{2!} & \\ & & \cdot & & \vdots \\ & & & \cdot & \frac{t^2}{2!} \\ & & & & t \\ & & & & 1 \end{bmatrix}$$

$$
= \begin{bmatrix} e^{t\lambda_i} & te^{t\lambda_i} & \cdot & & \frac{t^{r-1}e^{t\lambda_i}}{(r-1)!} \\ & e^{t\lambda_i} & te^{t\lambda_i} & \cdot & \\ & & \cdot & \cdot & \\ & & & \cdot & \\ & & & & te^{\lambda_i} \\ & & & \cdot & e^{t\lambda_i} \end{bmatrix} \tag{18}
$$

If $P = [p_{ij}]$ the fundamental matrix Pe^{tJ} has the form

$$
\begin{bmatrix} p_{11} & p_{12} & \cdots & p_{1n} \\ p_{21} & p_{22} & & p_{2n} \\ & & & \\ p_{n1} & p_{n2} & & p_{nn} \end{bmatrix} \begin{bmatrix} e^{tJ_1} & & & \\ & e^{tJ_2} & & \\ & & & \\ & & & e^{tJ_s} \end{bmatrix}
$$

where each e^{tJ_i} $(i = 1, \ldots, s)$ has the form given in (18). For definiteness, assume that the 1×1 matrices among the J_i are $J_1, \ldots, J_k$. Then if J_{k+1} is an $r \times r$ matrix, the columns of Pe^{tJ} are:

First Column	Second Column	$\cdots$	k-th Column	$(k+1)$-th Column
$p_{11}e^{t\lambda_1}$	$p_{12}e^{t\lambda_2}$		$p_{1k}e^{t\lambda_k}$	$p_{1,k+1}e^{t\lambda_{k+1}}$
$p_{21}e^{t\lambda_1}$	$p_{22}e^{t\lambda_2}$		$p_{2k}e^{t\lambda_k}$	$p_{2,k+1}e^{t\lambda_{k+1}}$
$\vdots$	$\vdots$		$\vdots$	$\vdots$
$p_{n1}e^{t\lambda_1}$	$p_{n2}e^{t\lambda_2}$		$p_{nk}e^{t\lambda_k}$	$p_{n,k+1}e^{t\lambda_{k+1}}$

$(k+2)$-th Column

$(tp_{1,k+1} + p_{1,k+2})e^{t\lambda_{k+1}}$

$(tp_{2,k+1} + p_{2,k+2}e^{t\lambda_{k+1}}$

$\vdots$

$(tp_{n,k+1} + p_{2,k+2})e^{t\lambda_{k+1}}$

$(k+r)$-th Column

$\left[\frac{t^{r-1}}{(r-1)!}p_{1,k+1} + \frac{t^{r-2}}{(r-2)!}p_{1,k+2} \right.$

$\left. + \cdots + tp_{1,k+r-1} + p_{1,k+r} \right]e^{t\lambda_{k+1}}$

(Because of the bulkiness of the expressions, we have written only the first element of the $(k+r)$-th column.)

In applications of differential equations the case in which A is real is of particular importance, and we want to apply the Real Canonical Form Theorem to obtain real solutions. It is sufficient to study in detail the case in which there exists a real nonsingular matrix P such that

$$P^{-1}AP = \begin{bmatrix} \alpha & \beta & 1 & 0 & & & & & \\ -\beta & \alpha & 0 & 1 & & & & & \\ & & \alpha & \beta & 1 & 0 & & & \\ & & -\beta & \alpha & 0 & 1 & & & \\ & & & & \cdot & & & & \\ & & & & & \cdot & & & \\ & & & & & & \alpha & \beta & 1 & 0 \\ & & & & & & -\beta & \alpha & 0 & 1 \\ & & & & & & & & \alpha & \beta \\ & & & & & & & & -\beta & \alpha \end{bmatrix}$$

Let

$$B = \begin{bmatrix} \alpha & & & & 0 \\ & \alpha & & & \\ & & \cdot & & \\ & & & \cdot & \\ 0 & & & & \alpha \end{bmatrix}$$

$$C = \begin{bmatrix} 0 & \beta & & & \\ -\beta & 0 & & & \\ & & \ddots & & \\ & & & 0 & \beta \\ & & & -\beta & 0 \end{bmatrix}$$

$$D \doteq \begin{bmatrix} 0 & 0 & 1 & & & 0 \\ & & & \ddots & & \\ & & & & 1 & \\ & & & & & 0 \\ 0 & & & & & 0 \end{bmatrix}$$

Matrix B commutes with matrix $C + D$ and $CD = DC$. Hence

$$e^{tPAP^{-1}} = e^{tB}e^{tC}e^{tD}$$

From the definition of the exponential of a matrix, it follows that

$$e^{tB} = \begin{bmatrix} e^{\alpha t} & & & \\ & e^{\alpha t} & & \\ & & \ddots & \\ 0 & & & e^{\alpha t} \end{bmatrix}$$

and

$$e^{tD} = \begin{bmatrix} 1 & 0 & t & 0 & \frac{t^2}{2!} & \cdots & \\ & 1 & 0 & t & 0 & \frac{t^2}{2!} & \cdots \\ & & & \ddots & & & \\ & & & & & t & \\ & & & & & 0 & \\ & & & & & 1 \end{bmatrix}$$

where the entry in the upper right hand corner is 0 since A is an $n \times n$ matrix where n is even. In order to compute e^{tC} we remark first that

$$\begin{bmatrix} 0 & 1 \\ -1 & 0 \end{bmatrix}^2 = \begin{bmatrix} -1 & 0 \\ 0 & -1 \end{bmatrix}$$

$$\begin{bmatrix} 0 & 1 \\ -1 & 0 \end{bmatrix}^3 = \begin{bmatrix} 0 & -1 \\ 1 & 0 \end{bmatrix}$$

$$\begin{bmatrix} 0 & 1 \\ -1 & 0 \end{bmatrix}^4 = \begin{bmatrix} 1 & 0 \\ 0 & 1 \end{bmatrix}$$

From the series expansions for the sine and cosine functions, it follows that

$$e^{tC} = \begin{bmatrix} \cos \beta t & \sin \beta t & & & & & \\ -\sin \beta t & \cos \beta t & & & & & \\ & & \cos \beta t & \sin \beta t & & & \\ & & -\sin \beta t & \cos \beta t & & & \\ & & & & \ddots & & \\ & & & & & \cos \beta t & \sin \beta t \\ & & & & & -\sin \beta t & \cos \beta t \end{bmatrix}$$

Hence

$$e^{tP^{-1}AP} = \begin{bmatrix} e^{\alpha t}\cos \beta t & e^{\alpha t}\sin \beta t & & & & \\ -e^{\alpha t}\sin \beta t & e^{\alpha t}\cos \beta t & & & & \\ & & \ddots & & & \\ & & & e^{\alpha t}\cos \beta t & e^{\alpha t}\sin \beta t \\ & & & -e^{\alpha t}\sin \beta t & e^{\alpha t}\cos \beta t \end{bmatrix}$$

$$\times \begin{bmatrix} 1 & 0 & t & & 0 & \frac{t^2}{2!} & & \\ & \cdot & \cdot & & \cdot & & \ddots & \\ & & \cdot & & & & \frac{t^2}{2!} \\ & & & t & & \cdot & & \\ & & & & \cdot & \cdot & & 0 \\ & & \cdot & & & \cdot & & t \\ & & & & \cdot & & & 0 \\ & & & & & \cdot & & 1 \end{bmatrix}$$

or

$$
e^{tP^{-1}AP} = \begin{bmatrix}
e^{\alpha t}\cos\beta t & e^{\alpha t}\sin\beta t & te^{\alpha t}\cos\beta t & te^{\alpha t}\sin\beta t \\
-e^{\alpha t}\sin\beta t & e^{\alpha t}\cos\beta t & -te^{\alpha t}\sin\beta t & te^{\alpha t}\cos\beta t \\
& & & & \cdots \\
\cdot & \cdot & e^{\alpha t}\cos\beta t & e^{\alpha t}\cos\beta t \\
\cdot & \cdot & -e^{\alpha t}\sin\beta t & e^{\alpha t}\cos\beta t \\
\cdot & \cdot & \cdot \\
\cdot & \cdot & \cdot
\end{bmatrix}
$$

As pointed out before $Pe^{tP^{-1}AP} = Pe^{tJ}$ is a fundamental matrix. The treatment of arbitrary real A requires direct extensions of the procedure described above.

Finally, we point out an important case. Suppose that the equation

$$x = Ax \tag{19}$$

corresponds to a single n-th order equation

$$x^{(n)} + a_1 x^{(n-1)} + a_2 x^{(n-2)} + \cdots + a_{n-1}x' + a_n x = 0$$

where $a_1, a_2, \ldots, a_n$ are constants. Then (19) has the form

$$
\begin{bmatrix} x_1' \\ x_2' \\ \vdots \\ x_n' \end{bmatrix} =
\begin{bmatrix}
0 & 1 & 0 & \cdots & 0 \\
0 & 0 & 1 & \cdots & 0 \\
& & \cdot & \cdot \\
0 & 0 & 0 & \cdots & 1 \\
-a_n & -a_{n-1} & & \cdots & -a_1
\end{bmatrix}
\begin{bmatrix} x_1 \\ x_2 \\ \vdots \\ x_n \end{bmatrix}
$$

First it is easy to prove (see Exercise 8) that the characteristic equation of

$$
A = \begin{bmatrix}
0 & 1 & 0 & \cdots & 0 \\
0 & 0 & 1 & \cdots & 0 \\
& & \cdot & \cdot \\
0 & 0 & 0 & \cdots & 1 \\
-a_n & -a_{n-1} & & \cdots & -a_1
\end{bmatrix}
$$

is $\lambda^n + a_1\lambda^{n-1} + a_2\lambda^{n-2} + \cdots + a_{n-1}\lambda + a_n = 0$. If λ is an eigenvalue of A and the Jordan Canonical form of A is

$$
\begin{bmatrix}
J_1 \\
& J_2 \\
& & \ddots \\
& & & J_s
\end{bmatrix}
$$

then λ appears in just one of the matrices J_i. The reason for this is that the linear space of eigenvectors associated with λ is 1-dimensional. In order to prove this, suppose that

$$\begin{bmatrix} x_1 \\ x_2 \\ \vdots \\ x_n \end{bmatrix}$$

is an eigenvector associated with an eigenvalue λ of A. Then

$$\begin{bmatrix} \lambda & -1 & 0 & \cdot & \cdot & \cdot & 0 \\ 0 & \lambda & -1 & \cdot & \cdot & \cdot & 0 \\ & \cdot & \cdot & \cdot & & & \\ 0 & 0 & \cdot & \cdot & \cdot & \lambda & -1 \\ a_n & a_{n-1} & \cdot & \cdot & \cdot & \cdot & \lambda + a_1 \end{bmatrix} \begin{bmatrix} x_1 \\ x_2 \\ \vdots \\ x_n \end{bmatrix} = 0$$

Then

$$\lambda x_1 - x_2 = 0$$
$$\lambda x_2 - x_3 = 0$$
$$\cdots$$
$$\lambda x_{n-1} - x_n = 0$$
$$a_n x_1 + a_{n-1} x_2 + \cdots + (\lambda + a_1) x_n = 0$$

or

$$\lambda x_1 = x_2$$
$$\lambda x_2 = x_3$$
$$\cdots$$
$$\lambda x_{n-1} = x_n$$
$$a_n x_1 + a_{n-1} \lambda x_1 + \cdots + (\lambda + a_1) \lambda^{n-1} x_1 = 0$$

Since λ is an eigenvalue, the last equation is satisfied for all real x_1

and hence each eigenvector has the form

$$\begin{bmatrix} x_1 \\ \lambda x_1 \\ \lambda^2 x_1 \\ \vdots \\ \lambda^{n-1} x_1 \end{bmatrix}$$

Thus the linear space is 1-dimensional.

The fundamental matrices which we have just computed give at once important information about the behavior of the solutions. That is, we have:

Theorem 2.8 *Let $R(\lambda)$ denote the real part of λ. Then each solution of*

$$x' = Ax \tag{20}$$

approaches zero as $t \to \infty$ $[t \to -\infty]$ iff $R(\lambda) < 0[R(\lambda) > 0]$ for all the eigenvalues λ of A. Each solution of (20) is bounded on the set (a, ∞) where a is any fixed real number [the set$(-\infty, a)$ where a is any fixed real number) iff

1) $R(\lambda) \le 0[R(\lambda) \ge 0]$ *for all the eigenvalues λ of A;*

2) $R(\lambda) = 0$ *implies that in the Jordan Canonical form, the eigenvalue λ appears only in matrices J_i such that J_i is a 1×1 matrix.*

Proof. The proof follows from inspection of fundamental matrix Pe^{tJ}, where J is the Jordan canonical form, and application of L'Hospital's Rule.   □

Theorem 2.9 *Suppose $\lambda = i\mu \ne 0$ is a pure imaginary eigenvalue of A, and let m be the number of matrices J_i in the Jordan canonical form of A in which λ or $-\lambda$ occurs. Suppose L is the linear space of solutions of (minimal) period $\frac{2\pi}{\mu}$ of (20). Then the dimension of L is m.*

Proof. As before, let $P = [p_{i,j}]$. Let $J_{q_1}, \ldots, J_{q_m}$ be the matrices in J in which λ or $-\lambda$ appear. Suppose the first column of J_{q_i} is contained in the k_i-th column of J. Then the m columns

$$\begin{bmatrix} p_{1k_i} & e^{it\mu} \\ \vdots & \\ p_{nk_i} & e^{it\mu} \end{bmatrix} \qquad (i = 1, \ldots, m) \qquad\qquad (21 - \text{i})$$

of Pe^{tJ} are linearly independent solutions each of which has period $\frac{2\pi}{\mu}$. Also no other column of Pe^{tJ} has minimal period $\frac{2\pi}{\mu}$. To complete the proof, it is sufficient to show that no linear combination of the columns of Pe^{tJ} which are different from the columns (21-i) $i = 1, \ldots, m$ has minimal period $\frac{2\pi}{\mu}$. This will be left as an exercise (Exercise 11). □

Corollary 2.1 *Suppose $\lambda = i\mu$ is a pure imaginary eigenvalue of the real matrix A and let q be the number of matrices J_i in the real canonical form of A in which μ occurs. Suppose L is the linear space of solutions of (minimal) period $2\pi/\mu$ of (20). Then the dimension of L is $2q$.*

Homogeneous Linear Equations With Period Coefficients: Floquet Theory

Next we obtain some results for homogeneous systems with periodic coefficients which parallel the results already obtained for homogeneous systems with constant coefficients. As might be expected, the results are not as explicit as the theorems for systems with constant coefficients. However, as will be seen later, they are useful in the study of certain nonlinear systems. The theory to be described is called Floquet theory. For this work, we need a lemma about matrices which says roughly that a log function for nonsingular matrices can be defined.

Lemma 2.3 *If C is a constant nonsingular matrix, there exists a constant matrix K such that*

$$C = e^K$$

Remark. As would be expected, the matrix K is not unique. E.g., if

$$K_1 = \begin{bmatrix} 2n\pi i & & 0 \\ & \ddots & \\ 0 & & 2n\pi i \end{bmatrix}$$

then $e^{K_1} = I$, the identity. Since $K_1 K = K K_1$ then

$$e^{K+K_1} = e^K e^{K_1} = e^K = C$$

Proof. Since C is nonsingular,

$$\det C \neq 0$$

and it follows that all eigenvalues $\lambda_1, \ldots, \lambda_n$ are nonzero. By the Jordan Canonical Form Theorem, there exists a constant nonsingular matrix P such that

$$P^{-1} C P = J$$

where J is the Jordan canonical form. We prove the lemma first for J. We have

$$J = \begin{bmatrix} J_1 & & & \\ & \cdot & & \\ & & \cdot & \\ & & & \cdot \\ & & & & J_s \end{bmatrix}$$

where, for $i = 1, \ldots, s$, either J_i is a 1×1 matrix or

$$J_i = \lambda_k \left(I + \frac{D}{\lambda_i} \right)$$

and

$$D = \begin{bmatrix} 0 & 1 & & & \\ & \cdot & \cdot & & \\ & & \cdot & \cdot & \\ & & & \cdot & 1 \\ & & & & 0 \end{bmatrix}$$

In analogy with the familiar infinite series from calculus,

$$\log(1 + x) = x - \frac{x^2}{2} + \frac{x^3}{3} - \frac{x^4}{4} + \cdots \quad (x \text{ real}, |x| < 1)$$

we define

$$\log\left(I + \frac{D}{\lambda_i}\right) = \sum_{k=1}^{\infty} \frac{(-1)^{k+1}}{k} \left(\frac{D}{\lambda_i}\right)^k \tag{22}$$

If J_i is an $r_i \times r_i$ matrix, then if $k \geq r_i$,

$$D^k = 0 \tag{23}$$

Hence the series on the right-hand side of equation (22) has only a finite number of nonzero terms and thus always converges. Moreover, it follows, again from (23), that

$$\exp\left[\log\left(I + \frac{D}{\lambda_i}\right)\right]$$

is also a polynomial in D/λ_i. Next we show that

$$\exp\left[\log\left(I + \frac{D}{\lambda_i}\right)\right] = I + \frac{D}{\lambda_i}$$

Using the infinite series for $\log(1 + x)$ and e^x, we have: if x is real and $|x| < 1$, then

$$1 + x = \exp[\log(1 + x)] \tag{24}$$

$$= 1 + [\log(1 + x)] + \frac{[\log(1 + x)]^2}{2!} + \frac{[\log(1 + x)]^3}{3!} + \cdots$$

$$= 1 + \left(x - \frac{x^2}{2} + \frac{x^3}{3} - \cdots\right) + \frac{1}{2!}\left(x - \frac{x^2}{2!} + \frac{x^3}{3} - \cdots\right)^2$$

$$+ \frac{1}{3!}\left(x - \frac{x^2}{2} + \frac{x^3}{3} - \cdots\right)^3 + \cdots$$

By standard theorems from calculus, it follows that the terms of the right-hand side of (24) can be rearranged to obtain a power series $\sum_{n=0}^{\infty} a_n x^n$ which has the same value as the original expression. Thus we have: for all real x such that $|x| < 1$,

$$1 + x = \sum_{n=0}^{\infty} a_n x^n$$

Hence by the Identity Theorem for Power Series it follows that $a_0 = 1$, $a_1 = 1$, and $a_k = 0$ for $k > 1$. Since $\log(I + D/\lambda_i)$ is defined in terms of the same formal expansion as the series expansion for $\log(1+x)$, it follows that the corresponding coefficients in $\exp[\log(I+D/\lambda_i)]$ must have the same values, i.e., we must have

$$\exp\left[\log\left(I + \frac{D}{\lambda_i}\right)\right] = I + \frac{D}{\lambda_i}$$

Now if J_i is a 1×1 matrix, let

$$K_i = \log \lambda_i$$

(As pointed out earlier, each λ_i is nonzero. Hence $\log \lambda_i$ is defined. Note, however, that λ_i may be negative or complex, and hence $\log \lambda_i$ is, in general, complex-valued.) If J_i is an $r_i \times r_i$ matrix where $r_i > 1$, let

$$K_i = (\log \lambda_i)I + \log\left(I + \frac{D}{\lambda_i}\right)$$

where I is the $r_i \times r_i$ identity matrix. Then if $r_i > 1$,

$$\exp K_i = \exp[(\log \lambda_i)I] \exp\left[\log\left(I + \frac{D}{\lambda_i}\right)\right]$$

$$= (e^{\log \lambda_i}I)\left(I + \frac{D}{\lambda_i}\right)$$

$$= \lambda_i\left(I + \frac{D}{\lambda_i}\right)$$

$$= J_i$$

Then if K is the matrix

$$K = \begin{bmatrix} K_1 & & \\ & \ddots & \\ & & K_s \end{bmatrix}$$

we have

$$\exp(K) = J$$

In the general case

$$C = PJP^{-1} = P(\exp K)P^{-1} = \exp(PKP^{-1})$$

This completes the proof of Lemma 2.3.   □

    Using Lemma 2.3, we prove a theorem which makes possible some useful definitions.

Theorem 2.10 *Suppose that $X(t)$ is a fundamental matrix of*

$$x' = A(t)x$$

where $A(t)$ is continuous for all real t and A has period τ, i.e., for all t,

$$A(t + \tau) = A(t)$$

Then $Y(t) = X(t + \tau)$ is a fundamental matrix. Also there exists a continuous matrix $P(t)$ of period τ such that $P(t)$ is nonsingular for all t and a constant matrix R such that

$$X(t) = P(t)e^{tR}$$

Proof. First since $X(t)$ is a fundamental matrix and A has period τ, we have:

$$Y'(t) = X'(t + \tau) = A(t + \tau)X(t + \tau) = A(t)Y(t)$$

and for all real t,

$$\det Y(t) = \det X(t + \tau) \neq 0.$$

Thus $Y(t)$ is a fundamental matrix. Hence by Theorem 2.4, there exists a nonsingular constant matrix C such that

$$X(t + \tau) = Y(t) = X(t)C$$

By Lemma 2.3, there exists a constant matrix R such that $C = e^{\tau R}$. Let

$$P(t) = X(t)e^{-tR} \tag{25}$$

Then for all t, matrix $P(t)$ is nonsingular because $X(t)$, being a fundamental matrix, is nonsingular for all t and e^{-tR} has the inverse e^{tR}. Multiplying (25) on the right by e^{tR}, we obtain:

$$P(t)e^{tR} = X(t)$$

Also $P(t)$ has period τ because

$$
\begin{aligned}
P(t+\tau) &= X(t+\tau)e^{-(t+\tau)R} \\
&= X(t)e^{\tau R}e^{-(t+\tau)R} \\
&= X(t)e^{-tR} \\
&= P(t)
\end{aligned}
$$

This completes the proof of Theorem 2.10.   □

Definition. The characteristic roots or eigenvalues of $C = e^{\tau R}$ are the *characteristic multipliers* of $A(t)$. The characteristic roots or eigenvalues of R are the *characteristic exponents* of $A(t)$.

(Note that since $X(t+\tau) = X(t)C$, then $X(t)$ is "multiplied by" the eigenvalues of C to get $X(t+\tau)$. Hence the name "characteristic multipliers.")

Theorem 2.11 *The characteristic multipliers $\lambda_1, \ldots, \lambda_n$ are uniquely determined by $A(t)$ and all the characteristic multipliers are nonzero.*

Proof. If $X_1(t)$ is a fundamental matrix of

$$
x' = A(t)x
$$

then by Theorem 2.4 there exists a matrix C_1, constant and nonsingular, such that

$$
X = X_1C_1
$$

Hence

$$
X_1(t)C_1e^{\tau R} = X(t)e^{\tau R} = X(t+\tau) = X_1(t+\tau)C_1,
$$

and therefore

$$
X_1(t+\tau) = X_1(t)C_1e^{\tau R}C_1^{-1}
$$

Since $e^{\tau R}$ and $C_1e^{\tau R}C_1^{-1}$ are similar, then by Lemma 2.2 they have the same characteristic roots. If $\lambda_1, \ldots, \lambda_n$ are the characteristic multipliers, then since C is nonsingular,

$$
\left| \prod_{i=1}^{n} \lambda_i \right| = |\det C| \neq 0
$$

This completes the proof of Theorem 2.11.   □

Theorem 2.12 *If $\rho_1, \ldots, \rho_n$ are the characteristic exponents of $A(t)$, then the characteristic multipliers of $A(t)$ are $e^{\tau\rho_1}, \ldots, e^{\tau\rho_n}$. Also if ρ_i appears in boxes of dimensions $i_1, \ldots, i_q$ in the Jordan canonical form of R, then $e^{\tau\rho_i}$ appears in boxes of dimensions $i_1, \ldots, i_q$ in the Jordan canonical form of C; and conversely.*

Proof. Since $\rho_1, \ldots, \rho_n$ are the characteristic roots of R, then

$$\det[R - \rho_i I] = \det \begin{bmatrix} a_{11} - \rho_i & a_{12} & \cdot & \cdot & \cdot & a_{1n} \\ \cdot & \cdot & \cdot & \cdot & & \\ a_n & & \cdot & \cdot & \cdot & a_{nn} - \rho_i \end{bmatrix} = 0,$$
$$i = 1, \ldots, n$$

Hence

$$\det[\tau R - \tau\rho_i] = \tau^n \det \begin{bmatrix} a_{11} - \rho_i & a_{12} & \cdot & \cdot & \cdot & a_{1n} \\ \cdot & \cdot & \cdot & \cdot & & \\ a_{n1} & & \cdot & \cdot & \cdot & a_{nn} - \rho_i \end{bmatrix} = 0,$$
$$i = 1, \ldots, n$$

That is, $\tau\rho_1, \ldots, \tau\rho_n$ are the characteristic roots of τR. By the Jordan Canonical Form Theorem, there exists a constant nonsingular matrix P such that

$$P^{-1} R P = J$$

where J is the Jordan canonical form of R. Therefore

$$P^{-1}\tau R P = \tau J$$
$$\tau R = P\tau J P^{-1}$$
$$e^{\tau R} = e^{P\tau J P^{-1}} = P e^{\tau J} P^{-1} \tag{26}$$

By definition, the characteristic multipliers are the characteristic roots of $C = e^{\tau R}$. But (26) shows that $e^{\tau R}$ and $e^{\tau J}$ are similar and hence, by Lemma 2.2, have the same characteristic roots. Since

$$J = \begin{bmatrix} J_1 & & & \\ & J_2 & & \\ & & \ddots & \\ & & & J_s \end{bmatrix}$$

where each J_q $(q = 1, \ldots, s)$ has the form

$$J_q = \begin{bmatrix} \rho_q & 1 & & & \\ & \cdot & \cdot & \cdot & \\ & & \cdot & & \cdot \\ & & & \cdot & 1 \\ & & & & \rho_q \end{bmatrix}$$

then

$$\tau J = \begin{bmatrix} \tau J_1 & & & \\ & \tau J_2 & & \\ & & \ddots & \\ & & & \tau J_s \end{bmatrix}$$

where

$$\tau J_q = \begin{bmatrix} \tau\rho_q & \tau & \cdot & \cdot & \cdot & \\ & \tau\rho_q & \tau & & & \\ & & & \ddots & & \\ & & & \ddots & & \tau \\ & & & & & \tau\rho_q \end{bmatrix} \tag{27}$$

Also

$$e^{\tau J} = \begin{bmatrix} e^{\tau J_1} & & & \\ & e^{\tau J_2} & & \\ & & \ddots & \\ & & & e^{\tau J_s} \end{bmatrix} \tag{28}$$

From (27) and (28) it is clear that the characteristic roots of $e^{\tau J}$ are $e^{\tau\rho_1}, \ldots, e^{\tau\rho_q}$.

Finally, we must show that the boxes in which ρ_q appears in the canonical form of R have the same dimensions as the boxes in which $e^{\tau\rho_q}$ appears in the canonical form of C. But it can be shown (see Exercise 12) that the canonical form of $e^{\tau J}$ is

$$\begin{bmatrix} K_1 & & & \\ & K_2 & & \\ & & \ddots & \\ & & & K_s \end{bmatrix}$$

where for $q = 1, \ldots, s$

$$K_q = \begin{bmatrix} e^{\tau \rho_q} & 1 & & \\ & \ddots & & 1 \\ & & & e^{\tau \rho_q} \end{bmatrix}$$

and K_q is a square matrix with the same number of rows as the square matrix J_q. This completes the proof of Theorem 2.12. $\square$

Theorem 2.13 *Each solution of $x' = A(t)x$ approaches zero as $t \to \infty$ if and only if $|\lambda_i| < 1$ for each characteristic multiplier λ_i {if and only if $R(\rho_i) < 0$ for each characteristic exponent ρ_i.} Each solution of $x' = A(t)x$ is bounded if and only if:*

1. *$|\lambda_i| \leq 1$ for each characteristic multiplier λ_i {$R(\rho_i) \leq 0$ for each characteristic exponent ρ_i} and*

2. *If $|\lambda_i| = 1$, each J_i in the canonical form of*

$$C = [X(t)]^{-1} X(t + \tau)$$

in which λ_i appears is a 1×1 matrix.

Proof. The theorem is proved by showing that there exists a change of variables under which the equation $x = A(t)x$ becomes a linear homogeneous system with constant coefficients. Using the notation of Theorem 2.10, let

$$Z(t) = e^{tR}[X(t)]^{-1} = [P(t)]^{-1}$$

and define

$$v(t) = Z(t)x(t)$$

where $x(t)$ is a solution, i.e.,

$$\frac{d}{dx}x(t) = A(t)x(t)$$

Then $x(t) = [Z(t)]^{-1}v(t)$ and

$$\frac{d}{dt}(Z^{-1}v) = A(t)Z^{-1}v$$

and we have

$$\left(\frac{d}{dt}Z^{-1}\right)v + Z^{-1}\frac{dv}{dt} = AZ^{-1}v$$

$$Z^{-1}\frac{dv}{dt} = AZ^{-1}v - \left(\frac{d}{dt}Z^{-1}\right)v$$

$$\frac{dv}{dt} = (ZAZ^{-1})v - Z\left(\frac{d}{dt}Z^{-1}\right)v$$

$$= [e^{tR}X^{-1}AXe^{-tR}]v - e^{tR}X^{-1}\left[\frac{dX}{dt}e^{-tR} + X\frac{d}{dt}e^{-tR}\right]v$$

$$= e^{tR}X^{-1}[AX - X']e^{-tR}v - e^{tR}X^{-1}X(-R)e^{-tR}v$$

$$(29)$$

Since X is a fundamental matrix, then

$$X' = AX$$

Thus (29) becomes

$$\frac{dv}{dt} = Rv$$

(Note: It must be emphasized that this reduction procedure is generally not computationally practical because in order to find R, it is necessary to find the characteristic multipliers, i.e., the eigenvalues of $[X(t)]^{-1}[X(t+\tau)]$. Thus one must first determine a fundamental matrix.)

The proof of the theorem follows from Theorem 2.12, the application of Theorem 2.8 to the equation

$$\frac{dv}{dt} = Rv$$

and the fact (from Theorem 2.10) that $|P(t)|$ is bounded. $\square$

Definition. The *geometric multiplicity* of an eigenvalue λ of matrix A is the dimension of the linear subspace

$$L = \{x/Ax = \lambda x\}$$

Theorem 2.14 *If no characteristic multiplier equals one, the equation*

$$x' = A(t)x \tag{30}$$

has no nontrivial periodic solutions of period τ. (That is, the only periodic solution is the identically zero solution.) If one is a characteristic multiplier of geometric multiplicity m, then the dimension of the linear space L of solutions with period τ of (30) is m.

Proof. Returning to the proof of Theorem 2.10, we have:

$$X(t) = P(t)e^{tR} = P(t)e^{tSJS^{-1}} = P(t)Se^{tJ}S^{-1}$$

where S is a nonsingular matrix and J is the Jordan canonical form of R. Denoting $P(t)S$ by $P_1(t)$, we conclude that $P_1(t)e^{tJ}$ is a fundamental matrix. If no characteristic multiplier equals one, then no characteristic root of J equals 0 or $\frac{2p\pi i}{\tau}(p = \pm 1, \pm 2, \dots)$. Hence no column of e^{tJ} has period τ or period $\frac{\tau}{p}$. But $P_1(t)$ has period τ. Hence no column of $P_1(t)e^{tJ}$ has period τ. Finally we need to show that no linear combination of columns of $P_1(t)e^{tJ}$ has period τ. The existence of such a linear combination means that there exists a constant vector c such that

$$P_1(t)e^{tJ}c = x(t)$$

where $x(t)$ has period τ. Then

$$e^{tJ}c = [P_1(t)]^{-1}x(t)$$

Thus $e^{tJ}c$ has period τ. But because of the form of e^{tJ}, this is impossible.

Now suppose one is a characteristic multiplier of geometric multiplicity m. It follows that e^{tJ} has m columns of period τ because if

$$\lambda_i = e^{\tau \rho_i} = 1$$

then

$$\tau \rho_i = 2p\pi i \qquad (p = 0, \pm 1, \pm 2, \dots)$$

and

$$e^{t\rho_i} = e^{t\frac{2p\pi i}{\tau}}$$

Hence the fundamental matrix $P_1(t)e^{tJ}$ has m columns of period τ, and thus the dimension of $\mathcal{L}$ is greater than or equal to m.

Finally we show that the dimension of $\mathcal{L}$ is less than or equal to m. If $x(t)$ is a solution of period τ of (30), there is a constant vector c such that

$$P_1(t)e^{tJ}c = x(t)$$

Hence

$$e^{tJ}c = [P_1(t)]^{-1} x(t)$$

has period τ. The components of $e^{tJ}c$ have typically the form

$$e^{t\rho_k}\left[c_k + c_{k+1}t + \cdots + c_{k+w}\frac{t^w}{w!}\right] \tag{31}$$

where $c_k, c_{k+1}, \dots, c_{k+w}$ are the kth, $(k+1)$th, $\dots ,(k+w)$th components of vector c. If expression (31) is not identically zero, then it has period τ if and only if

$$\rho_k = \frac{2\pi n i}{\tau}$$

where $n = 0, \pm 1, \dots$, and

$$c_{k+1} = \cdots = c_{k+w} = 0$$

Therefore $[P_1(t)]^{-1}x(t)$ is a linear combination of columns of e^{tJ} which have period τ. But there are m such columns of e^{tJ}. Thus the dimension of $\mathcal{L}$ is less than or equal to m. This completes the proof of Theorem 2.14. $\square$

A further examination of the fundamental matrix $P_1(t)e^{tJ}$ shows that equation (30) may have other periodic solutions. For example, if there is a characteristic exponent ρ_j such that

$$\rho_j = \frac{2\pi i}{m\tau}$$

where m is a positive integer, then there is a column of $P_1(t)e^{tJ}$ which has period $m\tau$. Also (30) may have almost periodic solutions. For example, if there is a pure imaginary characteristic exponent $\rho_j = i\alpha$, where α is real such that τ and $\frac{2\pi}{\alpha}$ are not rationally related, i.e., there do not exist integers a and b such that

$$\tau = \frac{a}{b}\frac{2\pi}{\alpha}$$

then the column

$$P_1(t)\begin{bmatrix} 0 \\ \vdots \\ e^{t\rho_j} \\ \vdots \\ 0 \end{bmatrix}$$

of $P_1(t)e^{tJ}$ is almost periodic because $P_1(t)$ has period τ and $e^{t\rho_j}$ has period $\frac{2\pi}{\alpha}$.

Inhomogeneous Linear Equations

Next we obtain a useful and important formula for the solution of linear inhomogeneous equations: the variation of constants formula.

Theorem (Variation of Constants Formula) *Suppose the $n \times n$ matrix $A(t)$ and the n-vector $u(t)$ are continuous on (a,b). Let $t_0 \in (a,b)$. If $X(t)$ is a fundamental matrix of*

$$x' = A(t)x$$

then the solution $x(t)$ of

$$x' = A(t)x + u(t) \tag{32}$$

which satisfies the initial condition

$$x(t_0) = 0$$

is: for $t \in (a, b)$,

$$x(t) = X(t) \int_{t_0}^{t} \{X(s)\}^{-1} u(s) ds$$

Proof. The proof is straightforward verification by computation.

$$\frac{d}{dt}[x(t)] = X'(t) \int_{t_0}^{t} [X(s)]^{-1} u(s) ds + X(t)\{X(t)\}^{-1} u(t)$$

$$= A(t)X(t) \int_{t_0}^{t} [X(s)]^{-1} u(s) ds + u(t)$$

$$= A(t)x(t) + u(t)$$

Corollary. *The solution $x(t)$ of (32) which satisfies the initial condition*

$$x(t_0) = x_0$$

is

$$x(t) = y(t) + X(t) \int_{t_0}^{t} [X(s)]^{-1} u(s) ds$$

where $y(t)$ is the solution of

$$x' = A(t)x$$

which satisfies the initial condition: $y(t_0) = x_0$.

The idea behind the variation of constants formula is this: If c is a constant vector then

$$x(t) = X(t)c$$

is a solution of

$$x' = A(t)x$$

One tries to obtain a solution of (32) by replacing the constant vector c by a vector function $c(t)$. Suppose that

$$x(t) = X(t)c(t)$$

is a solution of (32). Then

$$x' = Ax + u$$
$$x' = [X']c + [X]c' = AXc + Xc' = Ax + Xc'$$

Thus

$$Xc' = u$$

or

$$c' = X^{-1}u$$

Hence if $c(t_0) = 0$

$$c(t) = \int_{t_0}^{t} [X(s)]^{-1}u(s)ds$$

Sturm-Liouville Theory

So far in this book, we have been studying the initial value problem for differential equations. The initial value problem is the fundamental problem for differential equations, but there is another problem, the boundary value problem, which is also of great importance. Boundary value problems have a long history: they arose very early in the study of partial differential equations and have played an important role ever since. At this juncture, we will give a brief account of the boundary value problem for an important class of linear equations, i.e., the Sturm-Liouville theory. Although the results we will describe are very old (the original papers by Sturm and Liouville appeared in the 1830's) they remain important today, both in themselves and because they form the prototype for an important segment of linear functional analysis. Our account will be brief and uneven: brief because we will consider a special case and will omit proofs of serious theorems; uneven because we will go into detail at certain points where we want to explain connections between various results and, on the other hand, we will omit careful considerations of important topics.

It is natural to ask why these acknowledged omissions occur. A complete self-contained account would seem more useful. However, as will be seen from our summary of the Sturm-Liouville theory, the theory of boundary value problems is intimately dependent on functional analysis. As will be described, the Sturm-Liouville theory can be regarded as an infinite-dimensional analog of some matrix theory (linear algebra) which takes place in the Hilbert space $\mathcal{L}^2$. A complete exposition of the Sturm-Liouville theory should have functional analysis for its background. Since such a background requirement is contrary to the spirit of this book, we will settle for a lesser discussion.

First we describe how boundary value problems arise by looking at the heat equation. Consider a bar of uniform cross-section, homogeneous material and with its sides insulated so that no heat passes through them. Assume that the cross-section is small enough so that the temperature on any cross-section is constant. Let ℓ be the length of the bar. Then the temperature u at any point on the bar depends only on the position x on the bar and on the time t. That is, $u = u(x,t)$. The temperature $u(x,t)$ is prescribed by the heat equation

$$a^2 u_{xx} = u_t \tag{1}$$

where a is a positive constant depending on the material of the bar, and $0 < x < \ell$ and $t > 0$. (The endpoints $x = 0$, $x = \ell$ and $t = 0$ are not included because, for example, $u(x,t)$ is not defined for $x < 0$ and thus, strictly speaking, the derivative $u_x(0,t)$ is not defined. Only the right-hand derivative is defined at $x = 0$. As it turns out, this is not a serious complication in this work.) We assume that an initial temperature distribution is given in the bar, i.e.,

$$u(x,0) = f(x) \tag{2}$$

where $0 \leq x \leq \ell$ and $f(x)$ is a given function and we assume that the temperature at the ends of the bar is zero. That is,

$$u(0,t) = 0, \quad u(\ell,t) = 0 \tag{3}$$

for $t > 0$.

The objective is to solve equation (1) with initial condition (2) and boundary condition (3). We use the conventional approach of separation of variables and assume that

$$u(x,t) = X(x)T(t) \tag{4}$$

Substituting from (4) into (1), we obtain:

$$\alpha^2 X''T = XT'$$

or

$$\frac{X''}{X'} = \frac{1}{\alpha^2}\frac{T'}{T} \tag{5}$$

Since the left-hand side of (5) is a function of x only and the right-hand side is a function of t only it follows that both sides of (5) are equal to a constant k and thus solving (1) is reduced to solving the two ordinary differential equations

$$\frac{X''}{X} = k, \quad \frac{1}{\alpha^2}\frac{T'}{T} = k \tag{6}$$

The constant k remains unspecified. Applying the boundary conditions (3), we have:

$$X(0)T(t) = 0, \quad X(\ell)T(t) = 0 \tag{7}$$

Conditions (7) imply:

$$X(0) = 0, \quad X(\ell) = 0 \tag{8}$$

unless $T(t) = 0$ for all t. But we exclude the possibility that $T(t) = 0$ for all t because this would imply that

$$u(x,t) = 0$$

for all x and t. Then condition (2) could not be fulfilled except for the case in which $f(x) = 0$ for $0 \le x \le \ell$.

Thus we must solve the equation

$$X'' - kX = 0 \tag{9}$$

subject to the boundary conditions (8). This problem clearly has the trivial solution $X(x) = 0$ for all x. *The essential point to be made here is that the problem may or may not have nontrivial solutions. What happens depends on the value of k.* (Notice the radical difference between this problem and the initial value problem. The initial value problem has, under reasonable hypothesis on the differential equation, a unique solution. The boundary value problem may or may not have a solution.)

Suppose that k is a positive number, say $k = m^2$ where $m > 0$. Then the general solution of (9) is

$$X(x) = e_1 e^{mx} + c_2 e^{-mx}$$

and it follows that the problem described by (8) and (9) has no nontrivial solutions. Further calculations show that if $k = 0$ or if k is a complex number then the problem has no nontrivial solutions. (See, e.g., Boyce and DiPrima, Chapter 10.) However if k is a negative number, say $k = -q^2$ where $q > 0$, then the general solution of (9) is

$$X(x) = c_1 \cos qt + c_2 \sin qt$$

and from (8) it follows that $c_1 = 0$ and that

$$q = \frac{n\pi}{\ell} \qquad (n = \pm 1, \pm 2, \dots)$$

Hence

$$X(x) = c \sin \frac{n\pi}{\ell} x$$

where c is any nonzero constant. (To complete the solution of our original problem, described by equations (1), (2) and (3), we would then return to the equation

$$\frac{1}{\alpha^2} \frac{T'}{T} = -q^2$$

which is easily solved.) With considerably more work, we can establish the remarkable fact that the solutions

$$\sin \frac{n\pi}{\ell} x \qquad (n = 1, 2, \dots)$$

of our problem can be used to describe the initial temperature distribution $f(x)$ if $f(x)$ is a sufficiently well-behaved function. (E.g., if f is continuous on $[0, \ell]$, $f(0) = f(\ell)$ and f has a continuous first derivative on $(0, \ell)$.) In fact, if

$$b_n = \frac{2}{\ell} \int_0^\ell f(x) \sin \frac{n\pi x}{\ell} dx \qquad n = 1, 2, \dots$$

then the infinite series

$$\sum_{j-1}^{\infty} b_n \sin \frac{n\pi x}{\ell} \tag{10}$$

converges uniformly on $[0, \ell]$ to the function $f(x)$. This "remarkable fact" is, of course, a special case of the basic theorem of Fourier series. (We get a sine series because our function $f(x)$ is defined only on $[0, \ell]$. Extending $f(x)$ to an odd function on $[-\ell, \ell]$ and applying the basic Fourier theorem to that odd function yields the expansion (10).) For an introduction to Fourier series, see Churchill and Brown [1987].

The Sturm-Liouville theory which we will describe may be regarded as a generalization of the material summarized above. We start with a class of ordinary differential equations which arise from separation of variables applied to a class of partial differential equations. (The equation (9) with boundary conditions (8) is a prototype of this class of ordinary differential equations.) From this class of equations we obtain generalized Fourier series.

The theory to be obtained can be regarded as an infinite-dimensional version of the eigenvalue theory in linear algebra. So we will first list some theorems from linear algebra. These will then serve as a kind of guide through the Sturm-Liouville theory.

Let $A = [a_{ij}]$ be an $n \times n$ matrix with real constant entries. We will assume that A is a symmetric matrix, i.e.,

$$A = A^T$$

where A^T is the transpose of A, i.e., $A^T = [b_{ij}]$ where $b_{ij} = a_{ji}$. It is easy to show that if A is symmetric then

$$(Ax, y) = (x, Ay)$$

where x, y are n-vectors and (x, y) is the usual inner product, i.e., if $x = (x_1, \ldots, x_n)$ and $y = (y_1, \ldots, y_n)$, then $(x, y) = \sum_{i=1}^n x_i y_i$. We have the following results concerning matrix A:

Theorem 1A *If the eigenvalues of A are counted with their algebraic multiplicities, then A has n eigenvalues. Each eigenvalue is real, and if λ_0 is an eigenvalue of multiplicity q, then corresponding to λ_0, there are q linearly independent eigenvectors.*

Theorem 2A *Suppose that $\lambda_1, \ldots, \lambda_m$ are the distinct eigenvalues of A (thus $m \leq n$) and λ_j has multiplicity r_j $(j = 1, \ldots, m)$ and suppose that $x_{(j)}^{(1)}, \ldots, x_{(j)}^{(r_j)}$ is a set of linearly independent eigenvectors of λ_j $(j = 1, \ldots, m)$. Then the set of all the eigenvectors, i.e., the set*

$$S = \left\{ x_{(1)}^{(1)}, \ldots, x_{(1)}^{(r_1)}, x_{(2)}^{(1)}, \ldots, x_{(2)}^{(r_2)}, \ldots, x_{(m)}^{(1)}, \ldots, x_{(m)}^{(r_m)} \right\}$$

is a set of n linearly independent eigenvectors. (Thus S is a basis for real Euclidean n-space. That is, if y is a real n-vector, there is a unique linear combination of vectors in S which is equal to y.)

Theorem 3A *If λ is not an eigenvalue of A, then the vector equation*

$$(A - \lambda I)x = b$$

where b is a given n-vector has a unique solution x. (This is just a special case of Cramer's Rule since $\det(A - \lambda I) \neq 0$.)

Theorem 4A *If λ is an eigenvalue of A, then the vector equation*

$$(A - \lambda I)x = b \tag{11}$$

where b is given, has a solution if and only if for each vector y which is an eigenvector of A corresponding to eigenvalue λ it is true that

$$(b, y) = 0$$

Theorem 5A *If* (11) *has a solution, say* $\tilde{x}$, *then* (11) *has an infinite set of solutions of the form*

$$\tilde{x} + y$$

where y *is any element in the linear space of eigenvectors corresponding to eigenvalue* λ.

In describing the Sturm-Liouville theory, we work with real-valued functions on the interval $[0,1]$ and the inner product

$$(f,g) = \int_0^1 f(x)g(\bar{x})dx$$

where $\bar{g}(x)$ denotes the complex conjugate of $g(x)$. In a rigorous description this integral is a Lebesgue integral and function f,g are measurable functions which are elements of $\mathcal{L}^2[0,1]$, i.e., $\int_0^1 |f(x)|^2\, dx < \infty$ and $\int_0^1 |g(x)|^2\, dx < \infty$.

We study the differential equation

$$[p(x)y']' - q(x)y + \lambda y = 0 \tag{12}$$

where $p(x)$, $q(x)$ are continuous on $[0,1]$, and $p(x)$ is positive and differentiable on $[0,1]$ and λ is a number. We seek solutions $y(x)$ which satisfy the boundary conditions

$$ay(0) + by'(0) = 0 \tag{13}$$

$$cy(1) + dy'(1) = 0$$

where a, b, c, d are constants such that $a^2 + b^2 > 0$ and $c^2 + d^2 > 0$. The problem of finding numbers λ and corresponding nontrivial solutions of (12) which satisfy condition (13) is called a *Sturm-Liouville problem*. We shall refer to it as the $\mathcal{S}$-$\mathcal{L}$ problem. Let L denote the operator which takes the function u into

$$L(u) = -[p(x)u']' + q(x)u$$

It is clear, from the definition, that L is linear, i.e., if α, β are constants, then

$$L(\alpha u + \beta v) = \alpha L(u) + \beta L(v)$$

Simple calculations (see Boyce and DiPrima, Chapter 11) using integration by parts show that if u and v satisfy the boundary conditions (13) then

$$(L(u), v) - (u, L(v)) = 0 \qquad (14)$$

If L is a linear operator which satisfies (14), then L is said to be *self-adjoint*. (The self-adjointness of L is a generalization or extension of the symmetry property of matrix A, and it plays a crucial role in the development of the Sturm-Liouville theory.) The number λ is an *eigenvalue* of the $\mathcal{S}$-$\mathcal{L}$ problem if equation (12) has a nontrivial solution which satisfies the boundary conditions (13). A nontrivial solution is an *eigenfunction corresponding to eigenvalue λ.*

To anyone who encounters Sturm-Liouville theory for the first time, the form of equation (12) may seem strange and arbitrarily given. Actually the equation has neither of these properties. First it includes equations which arise in a large number of physical problems. Second, writing the equation in this form simplifies proofs that must be carried out, e.g., the proof of (14).

It is straightforward to show (see Boyce and DiPrima) that the eigenvalues (and eigenfunctions) of the $\mathcal{S}$-$\mathcal{L}$ problem are real, that each eigenvalue is simple (each eigenvalue has exactly one linearly independent eigenfunction)(see Exercise 13) and that linearly independent eigenfunctions (which must correspond to distinct eigenvalues) are orthogonal, i.e., if $\phi_1(x), \phi_2(x)$ are two such eigenvalues then

$$\int_0^1 \phi_1(x)\phi_2(x)dx = 0$$

A little more "machinery" mostly from complex variable theory can be used to prove that the set of eigenvalues is finite or is a denumerable set with no cluster points (see Coddington and Levinson, Chapter 7). Notice that we have not considered the question of whether there exist *any* eigenvalues. We have simply stated that *if* there exist eigenvalues and eigenfunctions they have the properties described above.

Now we indicate how to prove that the $\mathcal{S}$-$\mathcal{L}$ problem has an infinite set of eigenvalues. (We note first that our prototype $\mathcal{S}$-$\mathcal{L}$ problem, described by (8) and (9), has an infinite set of eigenvalues

$\frac{n^2\pi^2}{\ell^2}(n = 1, 2, \dots)$.) To prove that the $\mathcal{S}\text{-}\mathcal{L}$ problem has an infinite set of eigenvalues we introduce Green's function which we will define by using the variation of constants formula. Suppose that λ is not an eigenvalue of the $\mathcal{S}\text{-}\mathcal{L}$ problem. Let $\phi_1(t), \phi_2(t)$ be the linearly independent solutions of

$$L(u) + \lambda u = 0 \qquad (15)$$

which satisfy the conditions

$$\phi_1(0) = 1, \qquad \phi_1'(0) = 0$$

$$\phi_2(0) = 0, \qquad \phi_2'(0) = 1$$

(By the existence theorem in this chapter, we know that solutions ϕ_1 and ϕ_2 exists.)

If $f(t)$ is a continuous function on $[0, 1]$, then by the variation of constants formula (see Exercise 10, Chapter 2), a solution of

$$L(u) + \lambda u = f(t) \qquad (16)$$

is

$$u(t) = \int_0^t \left\{ \frac{\phi_1(t)\phi_2(s) - \phi_2(t)\phi_1(s)}{p(s)[W(\phi_1, \phi_2)(s)]} \right\} f(s)ds \qquad (17)$$

where $W(\phi_1, \phi_2)(s)$ is the Wronskian of ϕ_1 and ϕ_2 evaluated at s. We define the function $K_\lambda(t, s)$ as follows: if $s > t$, then

$$K_\lambda(t, s) = 0$$

if $s \leq t$, then

$$K_\lambda(t, s) = \frac{\phi_1(t)\phi_2(s) - \phi_2(t)\phi_1(s)}{p(s)[W(\phi_1, \phi_2)(s)]}$$

Then we may rewrite (17) as

$$u(t) = \int_0^1 \left\{ K_\lambda(t, s) \right\} f(s)ds \qquad (18)$$

Now let

$$(19) \qquad w(t) = \int_0^1 \left\{ K_\lambda(t, s) + k_1\phi_1(t) + k_2\phi_2(t) \right\} f(s)ds$$

where k_1, k_2 are functions of s. Since $u(t)$, defined in (17) and (18), is a solution of (16) and ϕ_1, ϕ_2 are solutions of the corresponding homogeneous equation(15), then it follows that $w(t)$ is a solution of (16). Also it can be proved that continuous functions $k_1(s), k_2(s)$ can be chosen so that, independent of f, $w(t)$ satisfies the boundary conditions (13). (See Coddington and Levinson, Chapter 7. For this proof, the condition that λ is not an eigenvalue of the $\mathcal{S}$-$\mathcal{L}$ problem is required.)

Equation (19) yields an analog of Theorem 3A. We have already shown that $w(t)$ is a solution of equation (16) with boundary conditions (13). The proof of uniqueness runs as follows: if $w_1(t)$ and $w_2(t)$ are solutions of (16) which satisfy boundary conditions (13), then

$$L(w_1 - w_2) + \lambda(w_1 - w_2) = 0$$

Also $w_1 - w_2$ satisfies boundary conditions (13). Since λ is not an eigenvalue of the $\mathcal{S}$-$\mathcal{L}$ problem, then for $t \in [0, 1]$,

$$w_1(t) - w_2(t) = 0$$

or

$$w_1(t) = w_2(t)$$

The function

$$G_\lambda(t, s) = K_\lambda(t, s) + k_1(s)\phi_1(t) + k_2(s)\phi_2(t)$$

is called the *Green's function* for the $\mathcal{S}$-$\mathcal{L}$ problem. We obtained Green's function by mathematical considerations starting from the variation of constants formula. For a different and interesting approach to Green's function in which one starts from a physical interpretation of equation (16), see Courant-Hilbert [1953] p. 351 ff.

Once we have Green's function, the set of eigenvalues is obtained from the following procedure. Without loss of generality, we may assume that $\lambda = 0$ is not an eigenvalue of the $\mathcal{S} - \mathcal{L}$ problem. Then $G_0(t, s)$ is defined. Let $\mathfrak{G}$ be the linear integral operator on $C[0, 1]$ defined by

$$\mathfrak{G}[f(t)] = \int_0^1 G_0(t, s)f(s)ds$$

First, from (19) we have: if $f(t)$ is continuous on $[0, 1]$, then

$$L\mathfrak{G}[f(t)] = L[w(t)] = f(t) \tag{20}$$

(We note that if u has a second derivative and satisfies the boundary conditions (13), then applying (19) with $f(s) = Lu$ to the identity

$$Lu = Lu$$

we obtain

$$u = \mathfrak{G}[L(u)] \tag{21}$$

Equations (20) and (21) show that $\mathfrak{G}$ is a kind of inverse of L.)

Now suppose λ is an eigenvalue of the $\mathcal{S}$-$\mathcal{L}$ problem and ϕ is an eigenfunction corresponding to λ. Then

$$L\phi = \lambda\phi$$

and

$$\phi = \int_0^1 G_0(s, t)\lambda\phi(s)ds$$

or

$$\phi = \lambda\mathfrak{G}(\phi) \tag{22}$$

That is, the operator $\mathfrak{G}$ has eigenvalue $\frac{1}{\lambda}$ and ϕ is an eigenfunction corresponding to that eigenvalue. (Note that it follows from (22) that $\lambda \neq 0$. For if $\lambda = 0$, then by (22), the function ϕ is identically zero.)

Conversely, suppose (22) holds. By (20), we have

$$\phi = L\mathfrak{G}(\phi)$$

and from (22)

$$L\mathfrak{G}(\phi) = L[\mathfrak{G}(\phi)] = L\left(\frac{\phi}{\lambda}\right)$$

Therefore

$$\phi = L\left(\frac{\phi}{\lambda}\right)$$

or

$$L(\phi) = \lambda\phi$$

Also it follows from (22) that since $\mathfrak{G}(\phi)$ satisfies the boundary conditions (13) then so does ϕ. Consequently λ is an eigenvalue and ϕ is a corresponding eigenfunction of the $\mathcal{S}$-$\mathcal{L}$ problem. Thus we have shown that the set of eigenvalues of the $\mathcal{S}$-$\mathcal{L}$ problem is the set of reciprocals of the eigenvalues of $\mathfrak{G}$. Hence it is sufficient to investigate the eigenvalues of $\mathfrak{G}$. It can be shown that the operator $\mathfrak{G}$ is self-adjoint; that $\mathfrak{G}$ is a completely continuous or compact operator from $C[0,1]$ into $C[0,1]$; and that $\mathfrak{G}$ has a denumerably infinite set of eigenvalues which have zero as a cluster point. (See Coddington and Levinson, Chapter 7.)

The arguments sketched above yield an infinite-dimensional result which is analogous to Theorem 1A stated earlier. The contrast between the two results is worth noting. To obtain the existence of eigenvalues of matrix A, we need only apply the fundamental theorem of algebra. Studying the eigenvalues of the $\mathcal{S}$-$\mathcal{L}$ problem requires considerably more effort.

Now we turn to an infinite-dimensional analog of Theorem 2A for the $\mathcal{S}$-$\mathcal{L}$ problem. Let $\{\lambda_m\}$ be the sequence of eigenvalues and let $\{\Psi_m(t)\}$ be a sequence of corresponding orthonormal eigenvectors. The analogous result (Coddington and Levinson, Chapter 7) is:

Theorem *If $f \in \mathcal{L}^2[0,1]$, then*

$$f = \sum_{m=0}^{\infty}(f, \Psi_m)\Psi_m$$

i.e.,

$$\lim_{k \to \infty} \left\| f - \sum_{m=0}^{k}(f, \Psi_m)\Psi_m \right\| = 0$$

(The series $\sum_{m=1}^{\infty}(f, \Psi_m)\Psi_m$ is a generalized Fourier series.)

Finally we consider equation (16) with boundary conditions (13) and assume that λ is an eigenvalue of our original $\mathcal{S}$-$\mathcal{L}$ problem (defined by (12) and (13)). We indicate how to obtain analogs of Theorems 4A and 5A by using some formal computations. (We emphasize that our argument will be only formal. To make the argument rigorous, we would need to justify each of the formal steps.)

By the theorem stated immediately above, we have

$$f = \sum_{j=1}^{\infty} (f, \Psi_j) \Psi_j$$

Assume that the desired solution u has the form

$$u = \sum_{j=1}^{\infty} \alpha_j \Psi_j$$

Then we must determine the α_j's.

Since

$$Lu = L\left\{ \sum_{j=1}^{\infty} \alpha_j \Psi_j \right\} = \sum_{j=1}^{\infty} \alpha_j L(\Psi_j) = \sum_{j=1}^{\infty} \alpha_j \lambda_j \Psi_j$$

then substituting in (16), we have

$$\sum_{j=1}^{\infty} \alpha_j \lambda_j \Psi_j = \sum_{j=1}^{\infty} \lambda \alpha_j \Psi_j + \sum_{j=1}^{\infty} (f, \Psi_j) \Psi_j$$

or

$$\sum_{j=1}^{\infty} \{\alpha_j \lambda_j - \lambda \alpha_j - (f, \Psi_j)\} \Psi_j = 0 \qquad (23)$$

Since the Ψ_j's are orthogonal, it follows that each coefficient in (23) is zero. Thus

$$\alpha_j \lambda_j - \lambda \alpha_j = (f, \Psi_j) \qquad (j = 1, 2, \ldots) \qquad (24)$$

By hypothesis, λ is an eigenvalue, say $\lambda = \lambda_m$. Then if $j \neq m$, we have

$$\alpha_j (\lambda_j - \lambda) = (f, \Psi_j)$$

and

$$\alpha_j = \frac{(f, \Psi_j)}{\lambda_j - \lambda}$$

But if $j = m$, then (24) becomes:

$$0 = (f, \Psi_m)$$

Thus if the problem defined by (16) and (13) has a solution, we must have:

(25) $(f, \Psi_m) = 0$

Conversely if (25) holds, then the problem has the solution

$$u = \sum_{j=1}^{\infty} \alpha_j \Psi_j$$

where, if $j \neq m$, we have

$$\alpha_j = \frac{(f, \Psi_j)}{\lambda_j - \lambda}$$

If $j = m$, we have:

$$\alpha_m (\lambda_m - \lambda_m) = (f, \Psi_m)$$

Since $\lambda_m - \lambda_m = 0$ and $(f, \Psi_m) = 0$, there is no condition on α_m and hence equation (16) with boundary conditions (13) has an infinite set of solutions of the form

$$\sum_{j=1}^{m-1} \alpha_j \Psi_j + \sum_{j=m+1}^{\infty} \alpha_j \Psi_j + c \Psi_m$$

where c is any constant.

The Sturm-Liouville problem we have discussed is an especially simple case, and we want to indicate how this problem should be generalized in order to obtain a more complete treatment of Sturm-Liouville theory. Among the extensions of the theory which can be made are the following. First in equation (12), the term λy can be replaced by $\lambda r(x)y$ where $r(x)$ is continuous and positive on $[0, 1]$. Then the inner product becomes:

$$(f, g) = \int_0^1 f(x)g(x)r(x)dx$$

For simplicity we have restricted ourselves to initial conditions (13) which are called separated conditions, i.e., the first condition refers to values at $x = 0$ and the second equation to values at $x = 1$. (Cf. Exercise 14.) The theory may be extended to cases where $p(x)$ is zero at one or both of the endpoints of $[0, 1]$. Bessel's equation is an important example in which $p(x)$ is zero at both endpoints. See Courant-Hilbert, vol. 1. The theory can be generalizes to treatment of an nth-order differential equation (with appropriate boundary conditions). See Coddington and Levinson, Chapter 7. Finally the Sturm-Liouville theory can be used to initiate studies of nonlinear boundary condition problems.

Exercises for Chapter 2

1. Find the eigenvalues and corresponding eigenvectors and generalized eigenvectors of the matrix

$$A = \begin{bmatrix} 1 & 6 & -5 & 6 & -6 & 1 & -1 \\ -3 & 7 & -3 & 7 & -6 & 1 & -1 \\ 0 & 0 & 1 & 6 & -5 & 1 & -1 \\ 0 & 0 & -3 & 7 & -2 & 1 & -1 \\ 0 & 0 & 0 & 0 & 2 & 1 & -1 \\ 0 & 0 & 0 & 0 & 0 & 2 & 0 \\ 0 & 0 & 0 & 0 & 0 & 0 & 2 \end{bmatrix}$$

2. Find the Jordan canonical form and the real canonical form of the matrix A in Exercise 1.

3. Find a real fundamental matrix for the equation

$$\frac{dx}{dt} = Ax$$

where A is the matrix in Exercise 1.

4. Given the nth order differential equation

$$\frac{d^n x}{dt^n} + a_1(t)\frac{d^{n-1}x}{dt^{n-1}} + \cdots + a_{n-1}(t)\frac{dx}{dt} + a_n(t)x = 0 \quad (1)$$

where $a_i(t)$ is continuous on (t_1, t_2) for $i = 1, \ldots, n$; let $\phi_1(t), \ldots,$ $\phi_n(t)$ be solutions of (1) on $(t_1 t_2)$.

Definition. The determinant

$$W(\phi_1, \ldots, \phi_n) = \det \begin{bmatrix} \phi_1 & \cdots & \phi_n \\ \frac{d\phi_1}{dt} & \cdots & \frac{d\phi_n}{dt} \\ \cdot & \cdot & \cdot \\ \frac{d^{n-1}\phi_1}{dt^{n-1}} & \cdots & \frac{d^{n-1}\phi_n}{dt^{n-1}} \end{bmatrix}$$

is the *Wronskian* of (1) with respect to $\phi_1, \ldots, \phi_n$.

Prove that a necessary and sufficient condition that $\phi_1, \ldots, \phi_n$ are linearly independent on (t_1, t_2) is that $W(\phi_1, \ldots, \phi_n) \neq 0$ for all $t \in (t_1, t_2)$ If t_0 and t are two points in (t_1, t_2) find a relationship between the values of the Wronskian at t_0 and t.

5. Suppose

$$P(t) = a_0 t^m + a_1 t^{m-1} + \cdots + a_{m-1}t + a_m$$
$$Q(t) = b_0 t^n + b_1 t^{n-1} + \cdots + b_{n-1}t + b_n$$

where $a_0 \neq 0$, $b_0 \neq 0$, and $m > n$. Prove that $P(t)$ and $Q(t)$, as functions on the real line, are linearly independent.

6. Let $f(t) = t^3$, $g(t) = t^4$. By Exercise 5, functions f and g are linearly independent as functions on the real line. Prove that f and g cannot be solutions of a differential equation of the form

(1) $$\frac{d^2 x}{dt^2} + a(t)\frac{dx}{dt} + b(t)x = 0$$

where $a(t)$, $b(t)$ are continuous for all real t.

7. Suppose $p(t)$ is a continuous real-valued function of period T. The equation

$$\frac{d^2x}{dt^2} + p(t)x = 0 \qquad (H)$$

is called Hill's equation.

Write Hill's equation as a two-dimensional first-order system and let

$$X(t) = \left[\begin{array}{cc} x_{11}(t) & x_{12}(t) \\ x_{21}(t) & x_{22}(t) \end{array} \right]$$

be a fundamental matrix of the system such that $X(0) = I$, the identity matrix. If

$$X(t + T) = X(t)C$$

where $C = [c_{ij}]$ is a constant nonsingular matrix, prove that $\det C = 1$. If

$$2A = c_{11} + c_{22}$$

show that if $A^2 < 1$, all the solutions of equation (H) are bounded. Show that if $A^2 > 1$, equation (H) has solutions which are not bounded.

8. Prove (by induction) that if

$$A = \left[\begin{array}{ccccc} 0 & 1 & 0 & \cdots & 0 \\ 0 & 0 & 1 & \cdots & 0 \\ & & \cdots & & \\ 0 & 0 & 0 & \cdots & 1 \\ -a_n & -a_{n-1} & & \cdots & -a_1 \end{array} \right]$$

then

$$\det[A - \lambda I] = (-1)^n(\lambda^n + a_1\lambda^{n-1} + a_2\lambda^{n-2} + \cdots + a_{n-1}\lambda + a_n).$$

9. Given the single n-th order equation

$$x^{(n)} + a_1 x^{(n-1)} + a_2 x^{(n-2)} + \cdots + a_{n-1}x' + a_n x = 0 \qquad (*)$$

show that if the characteristic equation is

$$\lambda^n + a_1\lambda^{n-1} + \cdots + a_{n-1}\lambda + a_n = (\lambda - \lambda_1)^r(\lambda - \lambda_2)^s = 0$$

where $r + s = n$, then all the solutions of $(*)$ are of the form

$$A_1e^{\lambda_1 t} + A_2te^{\lambda_1 t} + \ldots A_r t^{r-1}e^{\lambda_1 t} \tag{$**$}$$
$$+ B_1e^{\lambda_2 t} + B_2te^{\lambda_2 t} + \cdots + B_s t^{s-1}e^{\lambda_2 t}$$

where $A_1, \ldots, A_r, B_1, \ldots, B_s$ are constants.

10. Derive a variation of constants formula for

$$(*) \quad x^{(n)} + a_1(t)x^{(n-1)} + \cdots + a_{n-1}(t)x^{(1)} + a_n^{(t)}x = b(t)$$

where $a_1(t), a_2(t), \ldots, a_n(t)$ and $b(t)$ are continuous on (α, β) and $\alpha < 0 < \beta$, which yields the solution $x(t)$ of $(*)$ that satisfies the initial condition $x(0) = \bar{x}_0, x'(0) = \bar{x}_1, \ldots x^{(n-1)}(0) = \bar{x}_{n-1}$.

11. Complete the proof of Theorem 2.9.

12. Prove that the canonical form of $e^{\tau J}$ (in the proof of Theorem 2.12) is

$$\begin{bmatrix} K_1 & & & \\ & K_2 & & \\ & & \ddots & \\ & & & K_s \end{bmatrix}$$

where

$$K_q = \begin{bmatrix} e^{\tau \rho_q} & 1 & & \\ & e^{\tau \rho_q} & \vdots & \\ & \vdots & 1 & \\ & & e^{\tau \rho_q} \end{bmatrix}$$

13. Prove that each eigenvalue of the $\mathcal{S}$-$\mathcal{L}$ problem is simple.

14. It can be proved that the eigenvalues of the Sturm-Liouville problem

$$y'' + \lambda y = 0$$
$$y(-1) = y(1)$$
$$y'(-1) = y'(1)$$

are real. Show that the eigenvalues are a denumerable set and that corresponding to each nonzero eigenvalue, there are two linearly independent eigenfunctions. (This does not contradict the result on Exercise 13. The boundary conditions in this problem are not separated. So the result in Exercise 13 is not applicable.)

Chapter 3

Autonomous Systems

Although we obtained some explicit formulas for solutions of linear systems in Chapter 2, the results obtained suggest that there is very little chance of obtaining such formulas for more extensive classes of equations. For example, we obtained the explicit formula Pe^{tJ} for a fundamental matrix of a linear hmogeneous system with constant coefficients $x' = Ax$. However, even if we consider the slightly more general system $x' = A(t)x$ where $A(t)$ has period τ, the results obtained are no longer explicit formulas. Consequently, in studying larger classes of equations which include certain nonlinear equations, we must resign ourselves to obtaining limited information about the solutions. We will look for nonnumerical or "qualitative" properties of solutions. In this chapter we take the first steps in these qualitative studies by studying the qualitative properties of solutions of autonomous systems.

General Properties of Solutions of Autonomous Systems

Definition. A system of differential equations in which the independent variable does not appear explicitly, i.e., a system of the form

$$x' = f(x)$$

is an *autonomous* system.

127

Autonomous systems arise naturally in the study of conservative mechanical systems. For example, the equations arising in celestial mechanics are autonomous.

We consider the n-dimensional autonomous system

$$x_i' = f_i(x_1, \ldots, x_n) \qquad (i = 1, \ldots, n) \tag{1}$$

where for $i = 1, \ldots, n$, the domain of f_i is an open set D in R^n and f_i satisfies a local Lipschitz condition, i.e., for each point $(x_1, \ldots, x_n) \in D$, there is a neighborhood N of $(x_1, \ldots, x_n)$ such that in N, the functions $f_1, \ldots, f_n$ satisfy a Lipschitz condition. (As usual, the Lipschitz condition is invoked to insure uniqueness of solution.) We assume that all the solutions of (1) to be considered are maximal.

Definition. Let $(x_1(t), \ldots, x_n(t))$ be a solution of (1) such that not all the functions $x_1(t), \ldots, x_n(t)$ are constant functions. Let I be the domain of $(x_1(t), \ldots, x_n(t))$. The underlying point set of the solution, i.e., the set of points

$$C = \{(x_1(t), \ldots, x_n(t)) \mid t \in I\}$$

which is a curve in the intuitive sense, is an *orbit* of (1). If $n = 2$, the orbit is often called a *path* or *characteristic*.

(Notice that we do not use the term "curve" in a precise sense. We could approach the concept of orbit in this way, but it would be unnecessarily elaborate.)

Our first step is to obtain a fundamental result concerning orbits, a result which can be described intuitively as follows: If the open set D is the domain of f_i $(i = 1, \ldots, n)$ in (1), then each point of D is contained in exactly one orbit or one constant solution of (1). That is, the orbits and constant solutions cover D but they do not intersect one another. Also no orbit crosses itself. (If an orbit intersects itself, it is a simple closed curve.) These properties of orbits are intuitively reasonable, but rigorous proofs require some rather fussy steps which we carry out in Lemmas 3.1, 3.2, 3.3, 3.4 and Theorems 3.1 and 3.2. We emphasize that these results are *not* valid for nonautonomous systems. (See Exercises 1 and 2.) After obtaining this basic result we introduce some other concepts which will be useful later in the study of autonomous systems: Ω-limit sets, invariant sets, and minimal sets. Also we will discuss systems of the form

$$x' = ax + by$$
$$y' = cx + dy$$

and this gives us an introduction to the concept of stability.

Lemma 3.1 *If* $(x_1(t), \ldots, x_n(t))$ *is a solution of* (1) *with domain* I *and if* h *is a real number, then* $(x_1(t+h), \ldots, x_n(t+h))$ *is a solution of* (1) *with domain*

$$I_h = \{t - h/t \in I\}$$

Proof. Let

$$\bar{x}_i(t) = x_i(t+h) \qquad (i = 1, \ldots, n)$$

Then letting $s = t + h$, we have, for $i = 1, \ldots, n$,

$$\frac{d\bar{x}_i(t)}{dt} = \frac{dx_i}{ds}(s)\frac{ds}{dt} = \frac{dx_i}{ds}(s) = f_i[x_1(s), \ldots, x_n(s)]$$
$$= f_i[x_1(t+h), \ldots, x_n(t+h)]$$
$$= f_i[\bar{x}_1(t), \ldots, \bar{x}_n(t)]$$

This completes the proof of Lemma 1. $\quad\square$

(Notice that the argument used in the proof of Lemma 3.1 breaks down if (1) is nonautonomous, i.e., if

$$f_i = f_i(t, x_1, \ldots, x_n)$$

because then

$$\frac{d}{dt}\bar{x}_i(t) = \frac{d}{dt}x_i(t+h) = f_i[t+h, x_1(t+h), \ldots, x_n(t+h)]$$
$$= f_i[t+h, \bar{x}_1(t), \ldots, \bar{x}_n(t)])$$

Lemma 3.1 shows that there are many more solutions than orbits because if C is an orbit which is the underlying point set of the solution $(x_1(t), \ldots, x_n(t))$ with domain I, then C is also the underlying point set of the solution $(x_1(t+h), \ldots, x_n(t+h))$ with domain I_h.

Lemma 3.2 *If* $(x_1(t), \ldots, x_n(t))$ *is a solution of* (1) *with domain* I *and* $(\bar{x}_1(t), \ldots, \bar{x}_n(t))$ *is a solution of* (1) *with domain* $\bar{I}$ *and there exist* $t_o \in I$, $\bar{t}_0 \in \bar{I}$ *such that*

$$(x_1(t_0), \ldots, x_n(t_0)) = (\bar{x}_1(\bar{t}_0), \ldots, x_n(\bar{t}_0))$$

then for all $t \in \bar{I}$,

$$(x_1(t + t_0 - \bar{t}_0), \ldots, x_n(t + t_0 - \bar{t}_0)) = (\bar{x}_1(t), \ldots, \bar{x}_n(t)) \qquad (2)$$

Proof. By Lemma 3.1

$$(x_1(t + t_0 - \bar{t}_0), \ldots, x_n(t + t_0 - \bar{t}_0))$$

is a solution of (1) with domain

$$I_{t_0 - \bar{t}_0} = \{t - (t_0 - \bar{t}_0)/t \in I\}$$

But if $t = \bar{t}_0$

$$(x_1(t + t_0 - \bar{t}_0), \ldots, x_n(t + t_0 - \bar{t}_0))$$
$$= (x_1(t_0), \ldots, x_n(t_0)) = (\bar{x}_1(\bar{t}_0), \ldots, \bar{x}_n(\bar{t}_0))$$

By the uniqueness of solution property of (1) and the fact that the solutions are maximal, it follows that (2) holds for all $t \in \bar{I}$. This completes the proof of Lemma 3.2.   □

Lemma 3.2 shows that if orbit C is the underlying point set of a solution $(x_1(t), \ldots, x_n(t))$ then every solution of which C is the underlying point set has the form $(x_1(t+k), \ldots, x_n(t+k))$ where k is a real constant. Thus for every solution of which C is the underlying point set the direction of increasing t on C is the same. Hence C may be regarded as a directed curve.

Definition. If $(x_1^0, \ldots, x_n^0) \in D$ is such that for $i = 1, \ldots, n$

$$f_i(x_1^0, \ldots, x_n^0) = 0$$

then $(x_1^0, \ldots, x_n^0)$ is an *equilibrium point* (or *critical point* or *singular point*) of (1). Note that if $(x_1^0, \ldots, x_n^0)$ is an equilibrium point, and if for all real t,

$$x_i(t) = x_i^0 \qquad (i = 1, \ldots, n)$$

then $(x_1(t), \ldots, x_n(t))$ is a solution of (1).

Theorem 3.1 *If* $(x_1^0, \ldots, x_n^0) \in D$, *then either* $(x_1^0, \ldots, x_n^0)$ *is an equilibrium point of* (1) *or* $(x_1^0, \ldots, x_n^0)$ *is contained in exactly one orbit of* (1).

Proof. Suppose that $(x_1^0, \ldots, x_n^0)$ is not an equilibrium point of (1). By Existence Theorem 1.1, there exists a solution $(x_1(t), \ldots, x_n(t))$ of (1) which satisfies the initial condition

$$(x_1(t_0), \ldots, x_n(t_0)) = (x_1^0, \ldots, x_n^0)$$

for some t_0 in the domain of $(x_1(t), \ldots, x_n(t))$. Thus $(x_1^0, \ldots, x_n^0)$ is contained in at least one orbit of (1). Now suppose that $(x_1^0, \ldots, x_n^0)$ is contained in orbits C and $\bar{C}$ which are the underlying point set of solutions $(x_1(t), \ldots, x_n(t))$ and $(\bar{x}_1(t), \ldots, \bar{x}_n(t))$ with domains I and $\bar{I}$, respectively. Then there exist $t_0 \in I$, $\bar{t}_0 \in \bar{I}$ such that

$$(x_1^0, \ldots, x_n^0) = (x_1(t_0), \ldots, x_n(t_0)) = (\bar{x}_1(\bar{t}_0), \ldots, \bar{x}_n(\bar{t}_0))$$

Hence by Lemma 3.2

$$C = \bar{C}$$

This completes the proof of Theorem 3.1 □

Theorem 3.1 can be stated intuitively as: orbits do not intersect.

Lemma 3.3 *If* $(x_1(t), \ldots, x_n(t))$ *is a solution of* (1) *with domain* I *such that there exist* $t_1, t_2 \in I$, *where* $t_1 \neq t_2$, *and such that*

$$(x_1(t_1), \ldots, x_n(t_1)) = (x_1(t_2), \ldots, x_n(t_2))$$

then I *is the real t-axis and for all real* t,

$$x_i(t_1 + t) = x_i(t_2 + t) \qquad i = 1, \ldots, n \tag{3}$$

or

$$x_i(t + t_2 - t_1) = x_i(t) \qquad i = 1, \ldots, n$$

Proof. For definiteness assume that $t_1 < t_2$. Then

$$[t_1, t_2] \subset I$$

Let
$$E = \{s/s \geq 0 \text{ and } (3) \text{ holds for } 0 \leq t \leq s\}$$

Set E is nonempty because, by hypothesis, $0 \in E$. Let t_0 be the lub of E. Suppose first that $t_0 = 0$. By Lemma 3.1, $(x_1(t_1 + t), \ldots, x_n(t_1 + t))$ and $(x_1(t_2 + t), \ldots, x_n(t_2 + t))$ are solutions of (1) and by (3), these two solutions have the same value at $t = 0$. Hence by the uniqueness of solution, there exists $\delta > 0$ such that if $|t| < \delta$,

$$x_i(t_1 + t) = x_i(t_2 + t) \qquad i = 1, \ldots, n$$

This contradicts the condition that $t_0 = 0$.

Next suppose that t_0 is a positive number. Then $[t_1, t_2 + t_0] \subset I$ and hence $t_1 + t_0 \in I$. If $0 \leq t < t_0$

$$x_i(t_1 + t) = x_i(t_2 + t) \qquad i = 1, \ldots, n$$

Hence for $i = 1, \ldots, n$,

$$x_i(t_1 + t_0) = \lim_{t \uparrow t_0} x_i(t_1 + t) = \lim_{t \uparrow t_0} x_i(t_2 + t)$$

Hence the domain of $(x_1(t_2 + t), \ldots, x_n(t_2 + t))$ contains t_0. For if it did not contain t_0, then the domain could be extended to include t_0 by the same argument as used in the proof of Extension Theorem 1.3, and this would contradict the maximality which is assumed for all solutions discussed. Now $(x_1(t_1 + t_0 + t), \ldots, x_n(t_1 + t_0 + t))$ and $(x_1(t_2 + t_0 + t), \ldots, x_n(t_2 + t_0 + t))$ are both solutions of (1) by Lemma 3.1 and have the same value at $t = 0$. Hence by the uniqueness of solutions, there exists $\delta > 0$ such that if $|t| < \delta$,

$$x_i(t_1 + t_0 + t) = x_i(t_2 + t_0 + t) \qquad (i = 1, \ldots, n)$$

This contradicts the definition of t_0. Thus E is the set of nonnegative reals. A similar argument shows that (3) holds for $t \leq 0$.   □

Lemma 3.4 *If there exists a set of pairs $\{(t_\nu, h_\nu)\}$ where each h_ν is positive and such that*

$$x_i(t_\nu + h_\nu) = x_i(t_\nu) \qquad (i = 1, \ldots, n)$$

and if $(x_1(t), \ldots, x_n(t))$ *is a nonconstant solution (i.e., a solution which is not an equilibrium point), then there exists a minimal positive number h such that for all real t,*

$$x_i(t + h) = x_i(t) \qquad (i = 1, \ldots, n)$$

Proof. By Lemma 3.3

$$x_i(t + h_\nu) = x_i(t) \qquad (i = 1, \ldots, n) \tag{4}$$

for each ν and for all real t. Suppose there is a monotonic sequence h_{ν_n} which converges to zero. Let τ_1 and τ_2 be arbitrary numbers such that $\tau_1 < \tau_2$. For each h_{ν_n}, there is an integer k_n such that

$$\tau_1 + (k_n - 1)h_{\nu_n} \leq \tau_2 \leq \tau_1 + k_n h_{\nu_n}$$

By (4),

$$x_i(\tau_1) = x_i(\tau_1 + m h_{\nu_n}) \qquad (i = 1, \ldots, n)$$

for all integers m. Hence since

$$\lim_n (\tau_1 + k_n h_{\nu_n}) = \tau_2$$

we have

$$x_i(\tau_1) = x_i(\tau_2)$$

Since τ_1, τ_2 are arbitrary, then $x_1(t), \ldots, x_n(t)$ is an equilibrium point. Contradiction. Thus the $\underset{\nu}{\mathrm{glb}}\{h_\nu\}$ is positive. It follows at once that $h = \underset{\nu}{\mathrm{glb}}\{h_\nu\}$. $\qquad \square$

From Lemmas 3.3 and 3.4, we have at once

Theorem 3.2 *If $(x_1(t), \ldots, x_n(t))$ is a solution of (1) with domain I which is not an equilibrium point and if there exist $t_1, t_2 \in I$, where $t_1 \neq t_2$, and such that*

$$(x_1(t_1), \ldots, x_n(t_1)) = (x_1(t_2), \ldots, x_n(t))$$

then I is the positive t-axis and there is a minimal positive number h such that for all real t,

$$x_i(t + h) = x_i(t) \qquad (i = 1, \ldots, n)$$

Also the orbit underlying solution $x_1(t), \ldots, x_n(t)$ is a simple closed curve.

Theorem 3.2 can be stated intuitively as: if an orbit intersects itself, it is a simple closed curve.

Definition. The number h in the conclusion of Theorem 3.2 is the *period* of the solution $(x_1(t), \ldots, x_n(t))$.

Definition. Equilibrium point $(x_1^0, \ldots, x_n^0)$ of (1) is *isolated* if there exists a neighborhood N of $(x_1^0, \ldots, x_n^0)$ such that the only equilibrium point of (1) in N is $(x_1^0, \ldots, x_n^0)$.

Theorem 3.3 *Suppose $(x_1^0, \ldots, x_n^0) \in D$ is an equilibrium point of (1) and suppose that for $i = 1, \ldots, n$, the function f_i has continuous first derivatives in a neighborhood of $(x_1^0, \ldots, x_n^0)$. Let*

$$f_{ij} = \frac{\partial f_i}{\partial x_j}(x_1^0, \ldots, x_n^0).$$

Then if $\det[f_{ij}] \neq 0$, the point $(x_1^0, \ldots, x_n^0)$ is an isolated equilibrium point.

Proof. Since the mapping from R^n into R^n given by

$$x \to f(x)$$

is differentiable, there is a neighborhood N of $(x_1^0, \ldots, x_n^0)$ such that if $x = (x_1, \ldots, x_n) \in N$ and $x^0 = (x_1^0, \ldots, x_n^0)$, then

$$f(x) = [f_{ij}](x - x^0) + R(x)$$

where

$$\lim_{|x-x^0| \to 0} \frac{|R(x)|}{|x - x^0|} = 0 \tag{5}$$

Hence if $x \in N$ and $x - x^0 \neq 0$,

$$[f_{ij}]^{-1} f(x) = x - x^0 + [f_{ij}]^{-1} R(x) \tag{6}$$

But if $|x - x^0|$ is sufficiently small, then by (5),

$$\left| [f_{ij}]^{-1} \right| \, |R(x)| < \frac{1}{2} |x - x^0|$$

Thus if $|x - x^0|$ is sufficiently small by nonzero,

$$f(x) \neq 0$$

This completes the proof of Theorem 3.3. $\square$

Notation. If $S = x(t)$ denotes a solution of a system

$$x' = f(x) \tag{7}$$

the underlying point set of solution S, i.e., the orbit of S, will be denoted by $0(S)$ or $0[x(t)]$.

Definition. If $x(t)$ is a solution of the system

$$x' = f(x)$$

and if there exists a number t_0 such that $x(t)$ is defined for all $t \geq t_0$, then an ω-*limit point* of solution $x(t)$ is a point x^0 such that there exists a sequence of real numbers $\{t_n\}$ with

$$\lim_n t_n = \infty$$

and

$$\lim_n x(t_n) = x^0$$

If S denotes the solution $x(t)$, then the set of ω-limit points of S will be denoted by $\Omega(S)$ or $\Omega[x(t)]$.

Definition. Suppose that f is continuous on an open set $D \subset R^n$. A set $E \subset D$ is *invariant* if and only if for each point $x^0 \in E$, it is true that, if $x(t)$ is a solution of (7) such that for some t, $x(t) = x^0$, then $0[x(t)] \subset E$.

Notation. If A, B are subsets of R^n, let $d(A, B)$ denote the number

$$\underset{\substack{p \in A \\ q \in B}}{\mathrm{g}\ell\mathrm{b}} |p - q|$$

Theorem 3.4 *Suppose that f is continuous on an open set $D \subset R^n$ and suppose that $x(t)$ is a solution of (7) which is not an equilibrium point and is such that there exists a number t_0 so that $x(t)$ is defined for all $t \geq t_0$ and there exists a number $B > 0$ such that for all $t \geq t_0$,*

$$|x(t)| < B \tag{8}$$

Then $\Omega[x(t)]$ is nonempty, closed, connected and invariant. Also

$$\lim_{t \to \infty} d[x(t), \Omega[x(t)]] = 0$$

Proof.   That $\Omega[x(t)]$ is nonempty follows from (8) and the Weierstrass-Bolzano Theorem.

To show that $\Omega[x(t)]$ is closed, suppose that q is a limit point of $\Omega[x(t)]$. Then there exists a sequence $\{q_n\} \subset \Omega[x(t)]$ such that

$$\lim_n |q_n - q| = 0 \tag{9}$$

Since $q_n \in \Omega[x(t)]$, there exists $\{t_n\}$ such that $t_n \to \infty$ and

$$|x(t_n) - q_n| < \frac{1}{n} \tag{10}$$

From (9) and (10), it follows that

$$\lim_n x(t_n) = q$$

Thus $q \in \Omega[x(t)]$ and hence $\Omega[x(t)]$ is closed.

Now suppose $\Omega[x(t)]$ is not connected. Then there exist disjoint nonempty open sets U_1 and U_2 such that

$$\Omega[x(t)] \subset U_1 \cup U_2$$
$$\Omega[x(t)] \cap U_1 \neq \phi$$
$$\Omega[x(t)] \cap U_2 \neq \phi$$

Since $\Omega[x(t)]$ is closed and bounded, the sets

$$M = \Omega[x(t)] \cap U_1$$

and

$$N = \Omega[x(t)] \cap U_2$$

are closed and bounded. Hence since $M \cap N = \phi$, then $d(M, N) = \delta > 0$. Since $M \cup N = \Omega[x(t)]$, there exists a monotonic increasing sequence $\{t_n\}$ such that

$$\lim_n t_n = \infty$$

and

$$\left.\begin{array}{ll} d[x(t_n), M] < \frac{\delta}{4} & \text{for } n \text{ odd} \\ d[x(t_n), N] < \frac{\delta}{4} & \text{for } n \text{ even} \end{array}\right\} \tag{11}$$

Since $x(t)$ is a continuous function of t, then the functions

$$\left.\begin{array}{l} g(t) = d[x(t), M] \\ h(t) = d[x(t), N] \end{array}\right\} \tag{12}$$

are continuous. From (11), it follows that if $t = t_{2m-1}$,

$$g(t) - h(t) < 0$$

and if $t = t_{2m}$, then

$$g(t) - h(t) > 0$$

Hence there exists $t_m^{(1)}$ such that

$$t_m^{(1)} \in (t_{2m-1}, t_{2m})$$

and such that

$$g(t_m^{(1)}) - h(t_m^{(1)}) = 0 \tag{13}$$

But there is a point $\bar{x} \in \Omega[x(t)]$ such that a subsequence of $\{x(t_m^{(1)})\}$ converges to $\bar{x}$. Hence from (12) and (13), it follows that

$$d[\bar{x}, M] = d[\bar{x}, N] \tag{14}$$

Since $\bar{x} \in \Omega[x(t)]$, then $\bar{x} \in M \cup N$. If $\bar{x} \in M$, then

$$d[\bar{x}, M] = 0$$

and

$$d[\bar{x}, N] \geq \delta$$

This contradicts (14). The assumption that $\bar{x} \in N$ leads to a similar contradiction. Hence $\Omega[x(t)]$ is connected.

Next we prove that

$$\lim_{t \to \infty} d \{x(t), \Omega[x(t)]\} = 0$$

Suppose the statement is not true. Then there exists a number $r > 0$ and a sequence $\{t_n\}$ such that $t_n \to \infty$ and for all n,

$$d \{x(t_n), \Omega[x(t)]\} \geq r \tag{15}$$

But the sequence $\{x(t_n)\}$ is bounded and hence contains a subsequence which converges to a point p. But $p \in \Omega[x(t)]$. This contradicts (15).

Finally we prove that $\Omega[x(t)]$ is invariant. Suppose $\bar{x}(t)$ is a solution of (7) which is not an equilibrium point and which is such that for $t = \bar{t}$,

$$\bar{x}(t) \in \Omega[x(t)]$$

Let t_0 be a point in the domain of $\bar{x}(t)$. Since

$$\bar{x}(\bar{t}) \in \Omega[x(t)]$$

then there exists $\{t_n\}$ with $t_n \to \infty$ such that

$$\lim_n x(t_n) = \bar{x}(\bar{t})$$

Then if $\tau = t_0 - \bar{t}$, it follows from the continuity of the solution as a function of the initial value (Corollary 1.1) that

$$\lim_n x(t_n + \tau) = \bar{x}(\bar{t} + \tau) = \bar{x}(t_0)$$

This completes the proof of Theorem 3.4 □

Definition. A *minimal* set relative to equation (7) is a set E which is: 1) nonempty, 2) closed, 3) invariant; and is such that no proper subset of E has these three properties.

Theorem 3.5 *If E is a nonempty invariant compact set, then E contains a minimal set.*

Proof. Let $\{E_\nu\}$ be the collection of nonempty closed invariant subsets of E. Since $E \in \{E_\nu\}$, the collection is nonempty. let $O_\nu = E_\nu^c$ the complement in R^n of E_ν. The collection $\{O_\nu\}$ is nonempty and is partially ordered by inclusion. Hence by Zorn's Lemma, there is a maximal linearly ordered subset $\{E_{\nu_n}\}$ of $\{E_\nu\}$. Let $\tilde{E} = \bigcap_n E_{\nu_n}$. Then $\tilde{E} \neq \phi$ because $\{E_{\nu_n}\}$ has the finite intersection property and E is compact. Also $\tilde{E}$ is invariant because it is the intersection of invariant sets. Now suppose $\tilde{E}$ is not minimal. Then there exists a proper subset G of $\tilde{E}$ such that G is nonempty, closed and invariant. This contradicts the condition that $\{E_{\nu_n}\}$ is a maximal linearly ordered subset of $\{E_\nu\}$. This completes the proof of Theorem 3.5 $\square$

The material discussed in this section is studied from a very different viewpoint in the subject of topological dynamics or abstract dynamical systems. See Nemytskii and Stepanov [1960] or Gottschalk and Hedlund [1955]. In this subject, the basic properties of solutions of autonomous equations are studied from an abstract viewpoint. Instead of orbits in subsets of R^n, one considers continuous families of transformations of a metric space into itself and requires that the continuous families satisfy the same kinds of properties that the orbits have.

Orbits Near an Equilibrium Point: The Two-Dimensional Case

In the remainder of this chapter, we restrict ourselves to the orbits of two-dimensional systems, i.e., orbits which are curves in the xy-plane.

We first describe in detail the behavior of orbits near an isolated equilibrium point of a two-dimensional linear system. The results obtained, besides being beautiful and useful, are a kind of prototype for the general stability theory that will be studied later. The analysis used is quite elementary; nevertheless, it represents an important intellectual step. This study, originated by Poincaré, is the beginning of the qualitative theory of ordinary differential equations. Part of Poincaré's genius lay in his remarkable ability to perceive the right directions for study, that is, directions in which significant and extensive progress could be made. The study of orbits near equilibrium points is such a direction.

We study the two-dimensional linear homogeneous system

$$
\begin{aligned}
x' &= ax + by \\
y' &= cx + dy
\end{aligned}
\tag{16}
$$

where a, b, c, d are real constants. We assume that

$$
\det \begin{bmatrix} a & b \\ c & d \end{bmatrix} \neq 0
$$

and hence that $(0,0)$ is the only equilibrium point of (16). Let A denote the matrix

$$
\begin{bmatrix} a & b \\ c & d \end{bmatrix}
$$

If P is a real nonsingular matrix such that $P^{-I}AP = J$ where J is the real canonical form, then a fundamental matrix of (16) is Pe^{tJ}. Hence, except for the distortion introduced by multiplying by P, the orbits of (16) are the underlying point sets of curves described by linear combinations of the columns of e^{tJ}. We disregard this distortion, i.e., we assume that A is in canonical form. Let λ_1, λ_2 be the eigenvalues of A, i.e., suppose that λ_1, λ_2 are the roots of the characteristic equation of A:

$$
\lambda^2 - (a + d)\lambda + (ad - bc) = 0
$$

Case I: λ_1, λ_2 are real, unequal and have the same sign. Then

$$
A = \begin{bmatrix} \lambda_1 & 0 \\ 0 & \lambda_2 \end{bmatrix}
$$

and

$$
e^{tA} = \begin{bmatrix} e^{t\lambda_1} & 0 \\ 0 & e^{t\lambda_2} \end{bmatrix}
$$

and an arbitrary solution is

$$\begin{bmatrix} c_1 e^{t\lambda_1} \\ c_2 e^{t\lambda_2} \end{bmatrix}$$

where c_1, c_2 are real constants. Suppose first that $\lambda_1 < \lambda_2 < 0$. then if $c_1 = 0$, $c_2 > 0$, the corresponding orbit is the positive y-axis and is directed toward the origin. Similar orbits are obtained if $c_1 = 0$, $c_2 < 0$ or if $c_1 < 0$, $c_2 = 0$ or if $c_1 > 0$, $c_2 = 0$. If $c_1 \neq 0$, $c_2 \neq 0$

$$\lim_{t \to \infty} \frac{c_2 e^{\lambda_2 t}}{c_1 e^{\lambda_1 t}} = \lim_{t \to \infty} \frac{c_2}{c_1} e^{(\lambda_2 - \lambda_1)t} = \left(\text{sign} \frac{c_2}{c_1} \right) \infty$$

and

$$\lim_{t \to -\infty} \frac{c_2 e^{\lambda_2 t}}{c_1 e^{\lambda_1 t}} = 0$$

Typical orbits are sketched (with arrows indicating their direction) in Figure 1.

Definition. If λ_1, λ_2 are unequal and negative, the equilibrium point $(0,0)$ is a *stable node*.

If $\lambda_1 > \lambda_2 > 0$, then the typical orbits look the same as the orbits for the case $\lambda_1 < \lambda_2 < 0$ except that they are oppositely directed. See Figure 2.

Definition. If $\lambda_1 > \lambda_2 > 0$, the equilibrium point is an *unstable node*.

Case II. $\lambda_1 = \lambda_2 = \lambda$ and the matrix A is

$$A = \begin{bmatrix} \lambda & 0 \\ 0 & \lambda \end{bmatrix}$$

Then an arbitrary solution is

$$\begin{bmatrix} c_1 e^{t\lambda} \\ c_2 e^{t\lambda} \end{bmatrix}$$

If $c_1 \neq 0$, $c_2 \neq 0$, the corresponding orbit is the intersection of the line that passes through the origin and has slope c_2/c_1 with the interior of the quadrant which contains the point (c_1, c_2). If $c_1 > 0$ and $c_2 = 0$, the corresponding orbit is the positive x-axis. Similar orbits are obtained if $c_1 < 0$, $c_2 = 0$, or $c_1 = 0$, $c_2 > 0$ or $c_1 = 0$, $c_2 < 0$. If $\lambda < 0$, all the orbits approach $(0,0)$ as indicated in Figure 3. If $\lambda > 0$, the orbits are oppositely directed as indicted in Figure 4.

Definition. If $\lambda_1 = \lambda_2 = \lambda$ and the canonical form is diagonal, the equilibrium point is a *stable note* if $\lambda < 0$ and is an *unstable node* if $\lambda > 0$.

Case III. $\lambda_1 = \lambda_2 = \lambda$ and the matrix A is

$$A = \begin{bmatrix} \lambda & 1 \\ 0 & \lambda \end{bmatrix}$$

Then a fundamental matrix is

$$\begin{bmatrix} e^{t\lambda} & te^{t\lambda} \\ 0 & e^{t\lambda} \end{bmatrix}$$

and an arbitrary solution is

$$\begin{bmatrix} c_1 e^{t\lambda} + c_2 te^{t\lambda} \\ c_2 e^{t\lambda} \end{bmatrix}$$

If $c_1 > 0$, $c_2 = 0$ and $c_1 < 0$, $c_2 = 0$, the corresponding orbits are the positive and negative x-axes. If $c_2 \neq 0$ then

$$\lim_{t \to \infty} \frac{c_2 e^{t\lambda}}{c_1 e^{t\lambda} + c_2 te^{t\lambda}} = \lim_{t \to \infty} \frac{c_2}{c_1 + c_2 t} = 0$$

and

$$\lim_{t \to -\infty} \frac{c_2 e^{t\lambda}}{c_1 e^{t\lambda} + c_2 te^{t\lambda}} = 0$$

If $c_2 > 0$ [< 0], the orbit is in the upper [lower] half-plane. If $\lambda < 0$, the orbits approach $(0,0)$ and are sketched in Figure 5. If $\lambda > 0$, the orbits are oppositely directed as indicated in Figure 6.

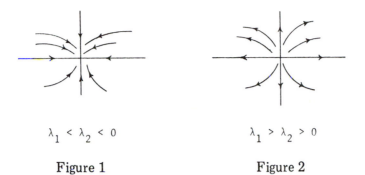

$$\lambda_1 < \lambda_2 < 0$$

Figure 1

$$\lambda_1 > \lambda_2 > 0$$

Figure 2

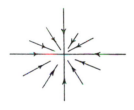

$$\lambda < 0$$

Figure 3

$$\lambda > 0$$

Figure 4

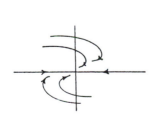

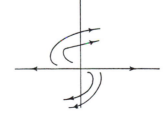

$$\lambda < 0$$

Figure 5

$$\lambda > 0$$

Figure 6

Definition. If $\lambda_1 = \lambda_2 = \lambda$ and the canonical form is

$$\begin{bmatrix} \lambda & 1 \\ 0 & \lambda \end{bmatrix}$$

the critical point is a *stable node* if $\lambda < 0$ and is an *unstable node* if $\lambda > 0$.

Case IV. $\lambda_2 < 0 < \lambda_1$. Then

$$A = \begin{bmatrix} \lambda_1 & 0 \\ 0 & \lambda_2 \end{bmatrix}$$

and an arbitrary solution is

$$\begin{bmatrix} c_1 e^{t\lambda_1} \\ c_2 e^{t\lambda_2} \end{bmatrix}$$

where c_1, c_2 are constants. If $c_1 > 0$, $c_2 > 0$, then both components are positive for all t. As $t \to \infty$, then $c_1 e^{t\lambda_1} \to \infty$ and $c_2 e^{t\lambda_2} \to 0$; as $t \to -\infty$, then $c_1 e^{t\lambda_1} \to 0$ and $c_2 e^{t\lambda_2} \to \infty$. The corresponding orbit is as sketched in the first quadrant in Figure 7. If c_1, c_2 are both negative or if one is positive and the other negative, then the corresponding orbits are in other quadrants but have similar properties and are also sketched in Figure 7. If $\lambda_2 = -\lambda_1$, the orbits are actually branches of hyperbolas. If $\lambda_2 \neq -\lambda_1$, then the orbit is a little more complicated: its equation is:

$$x \left(y^{-\frac{\lambda_1}{\lambda_2}} \right) = c_1 (c_2)^{-\frac{\lambda_1}{\lambda_2}}$$

Definition. If $\lambda_2 < 0 < \lambda_1$, the equilibrium point is a *saddle point*.

Case V. λ_1 and λ_2 are complex conjugate numbers, i.e., $\lambda_1 = \alpha + i\beta$, $\lambda_2 = \alpha - i\beta$.
    First if $\alpha = 0$ then an arbitrary solution is

$$\begin{bmatrix} c_1 \cos \beta t + c_2 \sin \beta t \\ -c_1 \sin \beta t + c_2 \cos \beta t \end{bmatrix} \tag{17}$$

where c_1, c_2 are real constants. Let ϕ be such that

$$\cos \phi = \frac{c_1}{\sqrt{c_1^2 + c_2^2}}$$

$$\sin \phi = \frac{c_2}{\sqrt{c_1^2 + c_2^2}}$$

and let

$$K = \sqrt{c_1^2 + c_2^2}$$

Then (17) can be written:

$$\begin{bmatrix} K \cos(\beta t - \phi) \\ -K \sin(\beta t - \phi) \end{bmatrix}$$

and the corresponding orbit is a circle with center $(0,0)$ and radius K. The orbits are sketched in Figure 8.

Definition. If $\lambda_1 = i\beta$, $\lambda_2 = -i\beta$, the equilibrium point is a *center*. If $\alpha \neq 0$, then an arbitrary solution is

$$\begin{bmatrix} c_1 e^{t\alpha} \cos \beta t + c_2 e^{t\alpha} \sin \beta t \\ -c_1 e^{t\alpha} \sin \beta t + c_2 e^{t\alpha} \cos \beta t \end{bmatrix}$$

or

$$\begin{bmatrix} K e^{t\alpha} \cos(\beta t - \phi) \\ -K e^{t\alpha} \sin(\beta t - \phi) \end{bmatrix}$$

and the corresponding orbit is a spiral which spirals outward [inward] if $\alpha > 0$ [< 0]. The orbits are sketched in Figures 9 and 10.

Definition. If $\lambda_1 = \alpha + i\beta$, $\lambda_2 = \alpha - i\beta$, where $\alpha \neq 0$, then the equilibrium point is a *spiral point* or *focus*. If $\alpha < 0$ [> 0], the equilibrium point is a *stable* [*unstable*] spiral point.

Now that the study of orbits near the equilibrium point $(0,0)$ of (16) is complete, it is rather natural to raise the following question. Suppose we consider the system

$$x' = ax + by + f(x, y)$$
$$y' = cx + dy + g(x, y) \tag{18}$$

where

$$\det \begin{bmatrix} a & b \\ c & d \end{bmatrix} \neq 0$$

and f and g are continuous, satisfy a local Lipschitz condition at each point in some neighborhood of $(0, 0)$, and are higher-order terms, i.e.,

$$\lim_{|x|+|y| \to 0} \frac{|f(x, y)| + |g(x, y)|}{|x| + |y|} = 0$$

Then $(0, 0)$ is an equilibrium point of (18) and the question is whether the behavior of the orbits of (18) in a small enough neighborhood of $(0, 0)$ is determined by the linear terms on the right-hand side of (18). Later we shall consider this question in a more general, n-dimensional context. For the present, we merely point out a result which is a special case of a theorem to be proved later. To state this result, we introduce the following definition.

Definition. The origin is an *asymptotically stable equilibrium point* of (16) [(18)] if given $\varepsilon > 0$ there exists $\delta > 0$ such that if $(x(t), y(t))$ is a solution of (16) [(18)] for which there is a number t_0 with

$$|x(t_0)| + |y(t_0)| < \delta$$

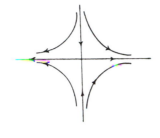

$$\lambda_2 < 0 < \lambda_1$$

Figure 7

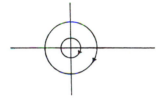

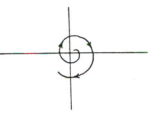

$\lambda_1 = i\beta$
$\lambda_2 = -i\beta$ $\beta > 0$

Figure 8

$\lambda_1 = \alpha + i\beta$
$\lambda_2 = \alpha - i\beta$ $\alpha > 0,\ \beta > 0$

Figure 9

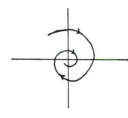

$\lambda_1 = \alpha + i\beta$
$\lambda_2 = \alpha - i\beta$ $\alpha < 0,\ \beta > 0$

Figure 10

then it is true that the solution $x(t)$, $y(t)$ is defined for all $t \geq t_0$, and

$$\operatorname*{lub}_{t \geq t_0} |x(t)| + |y(t)| < \varepsilon$$

and

$$\lim_{t \to \infty} x(t), y(t) = (0,0)$$

(If $(0,0)$ is a stable node or a stable spiral point of (16), then $(0,0)$ is clearly an asymptotically stable equilibrium point of (16).)

Later (Chapter 4) we will prove a more general version of the following theorem.

Theorem 3.6 *If $(0,0)$ is an asymptotically stable equilibrium point of (16), then $(0,0)$ is an asymptotically stable equilibrium point of (18).*

This theorem can be paraphrased roughly as: "If the behavior of the orbits of (16) near $(0,0)$ is not seriously affected by small changes, then the orbits near $(0,0)$ of (16) and (18) are about the same." But we are left then with the question of what happens if the orbits are seriously affected by small changes. For example, if $(0,0)$ is a center of (16) (and hence is not asymptotically stable), then a small change can seriously affect the orbits near $(0,0)$. It is easy to show that if $(0,0)$ is a center of (16), the addition of higher order terms (so that we have equation (18)) may make all the orbits move away from $(0,0)$ with increasing t or may make all the orbits move toward $(0,0)$ with increasing t. More precisely, motion with increasing t along all orbits is away from $(0,0)$ or motion with increasing t along all orbits is toward $(0,0)$.

For example, consider the system

$$x' = y + x^3$$
$$y' = -x + y^3$$

Let

$$r^2 = x^2 + y^2$$

Then

$$rr' = xx' + yy'$$
$$= xy + x^4 - xy + y^4$$
$$r' = \frac{x^4 + y^4}{r}$$

Thus for each solution other than the equilibrium point, r' is always positive. So each orbit moves away from $(0,0)$ as t increases. Thus $(0,0)$ is clearly not asymptotically stable.

On the other hand, for the system

$$x' = y - x^3$$
$$y' = -x - y^3$$

we have

$$r' = \frac{-x^4 - y^4}{r}$$

Thus r' is always negative, each orbit moves toward $(0,0)$ as t increases, and $(0,0)$ is asymptotically stable. One might say that the behavior of the orbits of the linear system

$$x' = y$$
$$y' = x$$

is not decisive and is strongly influenced by small or higher-order terms.

These examples show that, with the addition of certain nonlinear terms, the orbits all move away from $(0,0)$ or all move toward $(0,0)$. The behavior of the orbits may be more complicated, as the following example shows.

$$x' = y + xr^2 \sin \frac{\pi}{r}$$
$$y' = -x + yr^2 \sin \frac{\pi}{r} \tag{19}$$

where $r^2 = x^2 + y^2$. Then

$$rr' = xx' + yy'$$

$$= xy + x^2 r^2 \sin \frac{\pi}{r} - xy + y^2 r^2 \sin \frac{\pi}{r} = r^4 \sin \frac{\pi}{r}$$

and

$$r' = r^3 \sin \frac{\pi}{r}$$

Letting

$$\theta = \arctan \frac{y}{x}$$

we obtain:

$$\theta' = \frac{xy' - yx'}{x^2 + y^2}$$

$$= \frac{-x^2 + xyr^2 \sin \frac{\pi}{r} - y^2 - xyr^2 \sin \frac{\pi}{r}}{x^2 + y^2}$$

$$= \frac{-(x^2 + y^2)}{x^2 + y^2}$$

$$= -1$$

Thus

$$r = \frac{1}{n}$$

$$\theta = -t$$

describes an orbit which is a simple closed curve, i.e., a circle with center 0 and radius $1/n$. If $r > 1$, then $1/r < 1$ and hence $r' > 0$. Thus any orbit which passes through a point in the xy-plane that is outside the circle $r = 1$ moves away from the circle $r = 1$ with increasing t. Since $\theta' = -1$, the orbit spirals outward in the clockwise direction. Also, if

$$\frac{1}{2q} < r < \frac{1}{2q-1} \qquad (q = 1, 2, 3, \dots)$$

then

$$2q > \frac{1}{r} > 2q - 1$$

and hence

$$r' = r^3 \sin \frac{\pi}{r} < 0$$

If

$$\frac{1}{2q+1} < r < \frac{1}{2q} \qquad (q = 1, 2, 3, \dots,)$$

then

$$2q + 1 > \frac{1}{r} > 2q$$

and hence

$$r' > 0$$

Thus the orbits of (19) appear as sketched in Figure 11.

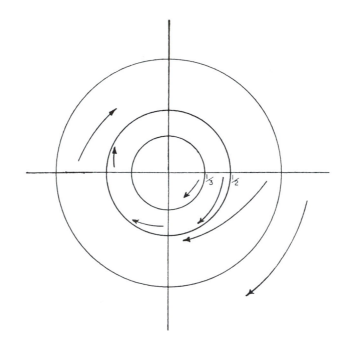

Figure 11

It can be proved that the kind of complication that occurs in this example cannot occur if a system of the form

$$x' = P(x, y)$$
$$y' = Q(x, y)$$

is studied where the functions P and Q are analytic, i.e., if P and Q can be represented by power series in x and y which converge in some neighborhood of $(0,0)$. (See Nemytskii and Stepanov [1960, Chapter II].)

Now we consider a somewhat more general question concerning orbits near an equilibrium point. Suppose that $P(x,y)$ and $Q(x,y)$ are power series in x and y which converge in a neighborhood N of $(0,0)$ and suppose that $(0,0)$ is an isolated equilibrium point of the system

$$x' = P(x, y)$$
$$y' = Q(x, y)$$

(20)

We say that $(0, 0)$ is a *general equilibrium point* of (20). Then what do the orbits of (20) in the neighborhood N look like? Theorem 3.6 gives a partial answer to this question because it says that under certain conditions, the linear terms in P and Q determine the behavior of the orbits. But considerably further analysis is needed. For example, so far we have no information about the behavior of the orbits if P or Q is a series which starts with terms of degree greater than one.

It might be thought that the answer to this question would be very complicated. There is, however, a beautiful and remarkably simple answer to this question which can be described roughly as follows. If $(0, 0)$ is an isolated equilibrium point of (20), then either $(0, 0)$ is a spiral, i.e., the orbits move toward $(0, 0)$ as t increases [decreases], or a center (there is a neighborhood N of $(0, 0)$ such that every orbit which has a nonempty intersection with N is a simple closed curve) or a neighborhood of $(0, 0)$ can be divided into a finite number of "sectors" of the following types:

I. Fan or parabolic sector

II. Hyperbolic sector

III. Elliptic sector

In each of these sketches, the arrows indicate the idrection of increasing time. Note that the only essentially new configuration that arises when nonlinear equilibrium points are studied is the elliptic sector.

For a precise statement of this result and a proof of it, see Lefschetz [Chapter X]. For a similar discussion see Nemytskii and Stepanov [1960, Chapter II].

A beautiful and sometimes useful concept that is introduced in the study of equilibrium points is the index of the equilibrium point. We will give here an intuitive version of the definition of index for equilibrium points in the two-dimensional case and leave to the appendix a rigorous definition for the n-dimensional case in terms of the notion of topological degree.

Definition. Let $P(x,y)$, $Q(x,y)$ be real-valued continuous functions on an open set D in the xy-plane. Let $V(x,y)$ be the vector field

$$V(x,y) = (P(x,y), Q(x,y))$$

i.e., V is a function with domain D and range contained in R^2. If $(x,y) \in D$, then $V(x,y)$ is the point or vector $(P(x,y), Q(x,y))$. An *equilibrium point* of $V(x,y)$ is a point $(x_0, y_0) \in D$ such that

$$P(x_0, y_0) = Q(x_0, y_0) = 0$$

Let C be a simple closed curve in D such that no point of C is an equilibrium point of V. We study how the vector associated with a point $(x_1, y_1) \in C$ changes as (x_1, y_1) is moved counterclockwise around C. The vector is rotated through an angle $j(2\pi)$ where j is an integer, positive or negative or zero. The number j is the *index of C with respect to $V(x,y)$*.

It is easy to see that this "definition" is not precisely formulted because several of the terms used have only an intuitive meaning. For example, what does it mean to "move counterclockwise" on C? As long as C is an easily visualized curve like a circle or ellipse, it is easy to describe precisely what we mean by "moving counterclockwise." But if C is a sufficiently "messy" curve, then it is not at all clear which is the counterclockwise direction on the curve. Secondly, although for easily visualized curves C, it is intuitively clear how to measure the angle through which the vector is rotated, no general and precise method for describing how to measure the angle has been given.

Thus our "definition" is far from rigorous. An efficient way to give a rigorous definition is to formultae the definition in terms of topological degree. This is done in the appendix of this book. The advantage of giving the definition in this way is that one obtains also a definition of index for the n-dimensional case where n is arbitrary. However, it should be pointed out that it is not difficult to give a rigorous definition of index for the 2-dimensional case which is much shorter than the definition for the n-dimensional case. Such a definition can also be obtained directly from the definition of winding number as given by Ahlfors [1966].

Having obtained the definition of the index of C with respect to $V(x, y)$, one can then prove that if C is continuously moved or deformed in such a way that it does not cross any equilibrium point of V during the continuous deformation, then the index j remains constant during the deformation. (In order to prove this result, one must first, of course, give a precise definition of "continuous deformation." This is done in the appendix.)

Now suppose that (x_0, y_0) is an isolated equilibrium point of V, i.e., suppose that there exists a circular neighborhood N of (x_0, y_0) such that there are no equilibrium points of V in N except (x_0, y_0). Let C_1, C_2 be simple closed curves in N such that (x_0, y_0) is in the interior of C_1 and is in the interior of C_2. It is not difficult to show that C_1 can be continuously deformed into C_2 without crossing any equilibrium point of V. Hence the index of C_1 equals the index of C_2, and indeed the index j is independent of the simple closed curve C provided C is contained in N and (x_0, y_0) is in the interior of C. Hence we may formulate the following definition.

Definition. The index j is the *index of the isolated equilibrium point* (x_0, y_0).

Notice that in obtaining this definition we have used the concept of the interior of a simple closed curve. This is a serious notion which requires careful definition. We discuss it in the appendix.

In the appendix, we shall see how to compute, on a rigorous basis, the indices of the equilibrium points at the origin of the various two-dimensional systems (16). However, using our intuitive definition of index of C with respect to V, it is easy to see, unrigorously, that the index of a node, stable or unstable, is $+1$ and the index of a saddle point is -1.

Finally we list, without proof. some important properties of the index.

1. The index of a simple closed curve which contains no equilibrium points in its interior is zero.

2. The index of a simple closed curve which has a finite set of equilibrium points in its interior is the sum of the indices of the equilibrium points.

3. Let C be a simple closed curve such that if $(x_1, y_1) \in C$, then the vector $V(x_1, y_1)$ is parallel to the tangent to C at (x_1, y_1). Then the index of C is ± 1. (In order to be able to speak of the tangent to C, we must, of course, assume that C is differentiable.)

4. Let C be a simple closed curve such that if $(x_1, y_1) \in C$ and if the vectors $V(x_1, y_1)$ point into the exterior of C, the index of C is $+1$.

5. The Bendixson formula: if (x_0, y_0) is an isolated equilibrium point of a 2-dimensional system

$$x' = P(x, y)$$
$$y' = Q(x, y)$$

then the index of (x_0, y_0) is $1 + (N_E - N_H)/2$ where N_E is the number of elliptic sectors and N_H is the number of hyperbolic sectors.

Unlike most of the previous discussion, our treatment of nonlinear two-dimensional systems is unrigorous and incomplete. The reason for this lies in the direction of study we wish to pursue. We are primarily concerned with applications of differential equation theory to biological and chemical problems. Most such problems require for their mathematical description an n-dimensional system where $n > 1$. Hence we are interested in developing theory which is valid for n-dimensional systems where $n > 2$.

In developing such theory in later chapters we will obtain as special cases some 2-dimensional results, e.g., Theorem 3.6 stated earlier in this section. On the other hand, some of the deep and beautiful results described here that are peculiar to the 2-dimensional case (e.g., the analysis of orbital behavior in the neighborhood of a general equilibrium point) seem unlikely to be of use in biological applications and hence have been very briefly dealt with.

Exercises for Chapter 3

1. Consider the system

$$
\begin{aligned}
\frac{dx}{dt} &= f(t, x, y) \\
\frac{dy}{dt} &= g(t, x, y)
\end{aligned}
\qquad (*)
$$

where x, y are scalars. Then a solution $(x(t), y(t))$ describes a curve in the xy-plane. Show, with an example, that two such curves may intersect one another.

2. Give an example of $(*)$ in Exercise 1 for which there is a solution which describes a curve which crosses itself.

3. Prove the following theorem.

   Theorem (Bendixson Criterion). *Given the system*

$$
\begin{aligned}
x' &= P(x, y) \\
y' &= Q(x, y)
\end{aligned}
\qquad (1)
$$

where P and Q have continuous first partial derivatives with respect to x and y at each point of the (x, y)-plane. Then if the function

$$\frac{\partial P}{\partial x} + \frac{\partial Q}{\partial y}$$

is nonzero at each point of the (x, y)-plane, system (1) has no nontrivial periodic solutions.

Hint: Use Green's Theorem.

4. Find the general solution of the system

$$\frac{dx}{dt} = y + x(1 - x^2 - y^2)$$

$$\frac{dy}{dt} = -x + y(1 - x^2 - y^2)$$

Show that the circle $x^2 + y^2 = 1$ is the orbit of a solution of the system.

Hint: Transform the system into polar coordinates.

One important step in the study of a nonlinear autonomous equation is to determine the equilibrium points and how solutions near them behave. We do this now and at the end of Chapter 4 for some of the examples described at the end of Chapter 1.

5. Show that the equilibrium points of the Volterra equations

$$x' = ax - Ax^2 - cxy$$
$$y' = -dy + exy \tag{V}$$

are

$$(0,0)$$

$$\left(\frac{a}{A},0\right)$$

$$\left(\frac{d}{e},\frac{a}{c}-\frac{A}{c}\frac{d}{e}\right)$$

6. Finding the equilibrium points of the Hodgkin-Huxley equation is a serious computation. Setting the right-hand sides of $(H-H)$ equal to zero, we obtain at once

$$m = m_\infty(V)$$
$$h = h_\infty(V)$$
$$n = n_\infty(V)$$

But when these functions of V are substituted into the right-hand side of the first equation in $(H-H)$, we obtain the following messy equation which must be solved for V.

$$\frac{I}{C} - \frac{1}{C}\Big\{\bar{g}_{N_a}[m_\infty(V)]^3[h(V)][(V-V_{N_a})$$
$$+ g_K[n_\infty(V)]^3(V-V_n) + \bar{g}_L(V-V_L)\Big\} = 0$$

In the study by FitzHugh (1969), it is assumed that h and n are constants h_0 and n_0, respectively, and that the phenomena are approximately described by the first two equations:

$$V' = \frac{I}{C} - \frac{1}{C}\Big[\bar{g}_{N_a} m^3 h_0(V-V_{N_a}) + \bar{g}_K n_0^4(V-V_K)$$
$$+ \bar{g}_L(V-V_L)\Big] \qquad (A)$$

$$m' = \frac{m_\infty(V) - m}{\tau_m(V)}$$

This assumption is based on the idea that since τ_h and τ_n are much larger than τ_m the system can be approximated by taking τ_h and τ_n to be "infinite." But even finding the equilibrium points of (A) is nontrivial. See Fitz Hugh [1969].

7. Show that there are exactly two equilibrium points in the first octant of the Field-Noyes equations:

$$x' = k_1 Ay - k_2 xy + k_3 Ax - 2k_4 x^2$$
$$y' = -k_1 Ay - k_2 xy + k_5 fz$$
$$z' = k_3 Ax - k_5 z$$

and determine their coordinates in terms of A, f, k_1, k_2, k_3, k_4, k_5. (One of the two equilibrium points is the origin.)

8. Show that the Goodwin equations have an equilibrium point in the set

$$P = \{(x_1, \ldots, x_n) / x_i \geq 0, \ i = 1, \ldots, n\}$$

Determine the coordinates of the equilibrium point if $\rho = 1$.

Chapter 4

Stability

Introduction

The material in the previous chapters is basic to all further study of differential equations. The topic of this chapter, stability, is certainly fundamental, but our emphasis and comparatively lengthy treatment are partly motivated by applications to problems in the physical world, especially biological problems. To some extent, by placing a strong emphasis on stability, we are choosing now a particular path in our study of differential equations.

The subject of stability can be approached from two viewpoints. The less important and less interesting viewpoint is that of pure mathematics. That is, a reasonable mathematical problem is to generalize or extend some of the results we obtained concerning the orbits of 2-dimensional linear homogeneous systems to orbits of nonlinear systems of dimension $n > 2$. It is easy to see that if we attempt as fine an analysis in the more general situation, the results become extremely complicated. The example starting on page 149 of Chapter 3 shows how complicated the results can become in the nonlinear 2-dimensional case, and if we consider even the linear problem in the n-dimensional case, where $n > 2$, the resuls become quite complicated (see Exercise 1). As already indicated at the end of Chapter 3, we are forced to ask for a more modest result than a detailed description of the orbits. Indeed, one of the few questions we can ask which has a reasonable and uncomplicated answer is: under what conditions do solutions approach the equilibrium point or stay close to the equilibrium point for all sufficiently large t? Actually, the question can be made somewhat more general. We can ask under what conditions solutions approach or stay close to a given solution? Thus our first step is to say precisely what we mean by "approach" or "stay close to." These are stability properties. Then we seek sufficient conditions that these stability properties hold.

Far more interesting and important is the approach to stability theory from the viewpoint of applications to problems in the physical world. If we assume that some physical system is described with a fair degree of accuracy by a system of ordinary differential equations, then the next question is: how are the solutions of the system of ordinary differential equations reflected in the actual behavior of the physical system? For example, suppose the system of equations has an equilibrium point, that is, a solution in which all the components are constants. This corresponds to a state of the physical system in which all the significant quantities are constant. If the system of differential equations has an equilibrium point, can we expect that the physical system will display corresponding behavior, i.e., are all significant quantities constant? It is fairly clear that we cannot expect such behavior from the physical system unless the equilibrium point of the system of differential equations has some additional properties. If, for example, all solutions "approach" the equilibrium point, then it would be reasonable to expect corresponding equilibrium behavior of the physical system. On the other hand, if all solutions "go away from" the equilibrium point, then it seems highly unlikely that the existence of the equilibrium point would be reflected by corresponding behavior of the physical system. Thus, we are led again to the question: under what conditions do solutions approach or stay close to a given solution?

Definition of Stability

Our first step is to formulate this question precisely. This formulation and the theory based on it are due to the great Russian mathematician Lyapunov.

Definition. Given the n-dimensional system

$$x' = f(t, x) \tag{E}$$

where f has domain D, an open set in (t, x)-space (i.e., Euclidean $(n+1)$-space) which includes the positive t-axis and f is continuous on D; suppose that the solution $x(t)$ of (E) is defined for all $t > \tau$. Then solution $x(t)$ is *stable (on the right) in the sense of Lyapunov* if there exists $t_0 > \tau$ such that, if $x(t_0) = x^0$ and if $x(t)$ is denoted by $x(t, t_0, x^0)$, the following conditions are satisfied:

1. There exists a positive constant b such that if

$$|x^1 - x^0| < b$$

   then the solution $x(t, t_0, x^1)$ of (E) is defined for all $t \geq t_0$;

2. Given $\varepsilon > 0$, then there exists $\delta > 0$, where $\delta = \delta(\varepsilon, f, t_0, x^0)$, i.e., δ depends on ε, f, t_0, x^0, such that $\delta \leq b$ and such that if

$$|x^1 - x^0| < \delta$$

   then for all $t \geq t_0$,

$$|x(t, t_0, x^1) - x(t, t_0, x^0)| < \varepsilon$$

   Solution $x(t, t_0, x^0)$ is *asymptotically stable (on the right) in the sense of Lyapunov* if conditions (1) and (2) hold and, in addition, we have:

3. There exists $\bar{\delta} > 0$, where $\bar{\delta} = \bar{\delta}(f, t_0, x^0)$, i.e., $\bar{\delta}$ depends on f, t_0, x^0, such that $\bar{\delta} < b$ and such that if

$$|x^1 - x^0| < \bar{\delta}$$

   then

$$\lim_{t \to \infty} |x(t, t_0, x^1) - x(t, t_0, x^0)| = 0$$

Definition. If conditions (1) and (2) [conditions (1), (2), and (3)] are satisfied by the solutions $x(t, t_0, x^1)$ in a given nonempty subset M of the solutions of (E), then $x(t, t_0, x^0)$ is *conditionally stable* [*asymptotically conditionally stable*] (*on the right*) *in the sense of Lyapunov*.

Definition. Solution $x(t)$ is *unstable* if it is not stable.

Remarks. 1. Condition (1) says roughly that if a solution gets close enough to $x(t, t_0, x^0)$, then the solution is defined for all sufficiently large t. Condition (2) says roughly that if a solution gets close enough to $x(t, t_0, x^0)$, then it stays close to $x(t, t_0, x^0)$ for all later t. It is easy to show (see Example 2 below) that conditions (1) and (2) do not imply (3). Also there are examples (see Cesari [1971]) for which conditions (1) and (3) are satisfied, but for which condition (2) is not satisfied. Thus if condition (1) holds, conditions (2) and (3) are independent.

2. In the definition of stability, the initial condition (t_0, x^0) seems to play a prominent role. However, under reasonable hypotheses, it can be shown (Exercise 2) that if there exists an initial condition (t_0, x^0) such that conditions (1) and (2) [conditions (1), (2), and (3)] in the definition of stability [asymptotic stability] are satisfied then if $\bar{t}_0 > t_0$ and $x(\bar{t}_0) = \bar{x}^0$, then $(\bar{t}_0, \bar{x}^0)$ is an initial condition for which conditions (1) and (2) [conditions (1), (2), and (3)] are satisfied. Also if $\bar{t}_0 < t_0$ and solution $x(t)$ is defined for $t \geq \bar{t}_0$, the same statement holds. Hence stability [asymptotic stability] does not depend on the point t_0.

3. Stability (on the left) and asymptotic stability (on the left) are defined in almost parallel ways. The only difference is that t is decreasing instead of increasing.

4. The word "stable" used in denoting the various kinds of equilibrium points of linear homogeneous 2-dimensional systems studied in Chapter 3 (e.g. stable node) is not used in the same sense as in the definition of stable given above. All the equilibrium points in Chapter 3 that were termed "stable" are asymptotically stable in the sense of the definition given above.

5. Since we will be using the definitions of stability and asymp-

totic stability given above most of the time, we will generally omit "(on the right) in the sense of Lyapunov."

6. There are many definitions of stability. (For more extensive accounts of the theory see Lefschetz [1962], and Hahn [1967].) In formulating a definition of stability, one seeks a concept which seems to agree to some extent with an intuitive picture of the physical situation and is at the same time a condition that can be verified in particular cases and can be used as the basis for a coherent mathematical theory. The Lyapunov concepts of stability have these properties and hence have been studied extensively. However, the physical interpretation of the theory is far from satisfactory. At the end of this chapter, we will point out some of the difficulties.

Examples

1. Among the equilibrium points of linear homogeneous 2-dimensional systems studied in Chapter 3, the stable [unstable] node and spiral are asymptotically stable on the right [left], the center is stable, but not asymptotically stable, and the saddle point is conditionally asymptotically stable. For other examples of conditional asymptotic stability, we can look at the notion of fan (or parabolic sector) introduced in Chapter 3. Suppose F is a parabolic sector in which all the solutions approach the equilibrium point, say $(0,0)$. Then $(0,0)$ is asymptotically conditionally stable where the set M is the fan F.

2. For the (scalar) equation

$$x' = 0$$

every solution has the form

$$x(t) = k$$

where k is a constant. Hence every solution is stable but not asymptotically stable.

3. The (scalar) equation

$$x' = x^2$$

is easily solved by separating variables as follows:

$$\frac{dx}{x^2} = dt$$

$$-\frac{1}{x} = t + C$$

If, for $t = 0$, we require that $x = x^0 > 0$, then

$$-\frac{1}{x^0} = C \quad \text{and}$$

$$-\frac{1}{x} = t - \frac{1}{x^0} = \frac{x^0 t - 1}{x^0}$$

or

$$x(t) = \frac{x^0}{1 - x^0 t}$$

Thus the solution is not defined for $t = 1/x^0$. Thus $x(t) \equiv 0$ is a solution which is not stable because condition (1) is not satisfied.

4. Finally, we consider the (scalar) equation

$$x' = 1 - x^2$$

and investigate the stability of the solution

$$x(t) = -1 \quad \text{for all} \quad t.$$

Separating variables, we may solve the equation as follows:

$$\frac{dx}{1 - x^2} = dt \quad \text{or}$$

$$\text{arctanh}\, x = t + C \quad \text{for} \quad x \in (-1, 1)$$

Hence if $x = x^0 \in (-1, 1)$ for $t = t_0$, we have:

$$\text{arctanh } x^0 = t_0 + C$$
$$C = \text{arctanh } x^0 - t_0$$

Therefore

$$x(t) = \tanh(t - t_0 + K)$$

where $K = \text{arctanh } x^0$, and

$$\lim_{t \to \infty} x(t) = +1$$

If $x > 1$ or $x < -1$, we have

$$\text{arccoth } x = t + \bar{C}$$

Hence if $x = x^0$, where $|x^0| > 1$, for $t = t_0$, we have

$$\bar{C} = \text{arccoth } x^0 - t_0$$

Therefore

$$x = \coth(t - t_0 + \bar{K})$$

where $\bar{K} = \text{arccoth } x^0$, and thus the solution $x(t) \equiv -1$ is not stable on the right because neither condition (1) nor condition (2) is satisfied. Note, however, that condition (1) is satisfied for every solution whose initial value x^0 is such that $x^0 \in (-1, 1)$.

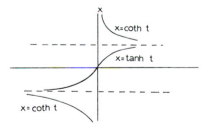

Stability of Solutions of Linear Systems

We begin with a remark about linear systems.

Theorem 4.1 *Given the linear system*

$$x' = A(t)x + f(t) \tag{1}$$

where $A(t)$ and $f(t)$ are continuous for each $t > t_0 - \delta$, where $\delta > 0$; then if there exists a solution $x(t, t_0, u^0)$ of (1) which is [asymptotically] stable then every solution of (1) is [asymptotically] stable.

Proof. Since $x(t, t_0, u^0)$ is stable, then $\varepsilon > 0$ implies there exists $\delta > 0$ such that if $|\Delta u| < \delta$, then for $t \geq t_0$

$$|x(t, t_0, u^0 + \Delta u) - x(t, t_0, u^0)| < \varepsilon \tag{2}$$

But

$$x(t, t_0, u^0 + \Delta u) - x(t, t_0, u^0)$$

and

$$x(t, t_0, u + \Delta u) - x(t, t_0, u)$$

are solutions of

$$x' = A(t)x$$

which both have value Δu at $t = t_0$. Hence by the uniqueness of solution and by (2), we have: for $t \geq t_0$

$$[x(t, t_0, u + \Delta u) - x(t, t_0, u)] = |x(t, t_0, u^0 + \Delta u) - x(t, t_0, u^0)| < \varepsilon$$

Thus $x(t, t_0, u)$ is stable. A slight extension of the argument shows that if $x(t, t_0, u^0)$ is asymptotically stable, $x(t, t_0, u)$ is asymptotically stable. $\square$

Now we state our main result.

Stability Theorem for Linear Systems. *Given the linear homogeneous system*

$$x' = A(t)x \tag{3}$$

where $A(t)$ is continuous for all $t > \bar{t}$, a fixed value, let 0 denote the identically zero solution of (3). Then the following conclusions hold.

(1) *If A is a constant matrix, then 0 is stable iff:*

    (i) *$R(\lambda) \leq 0$ for all eigenvalues λ of A, and*

    (ii) *if $R(\lambda) = 0$, then λ appears only in one-dimensional boxes in the Jordan canonical form of A.*

1. *If A is constant, then 0 is asymptotically stable iff $R(\lambda) < 0$ for all eigenvalues λ of A.*

2. *If $A(t)$ has period τ, then 0 is stable iff*

    (i) *$R(\rho) \leq 0$ for each characteristic exponent ρ of $A(t)$, and*

    (ii) *if $R(\rho) = 0$, then ρ appears only in one-dimensional boxes in the Jordan canonical form of R (where R is the matrix introduced in Theorem 2.10).*

3. *If $A(t)$ has period T, then 0 is asymptotically stable iff $R(\rho) < 0$ for each characteristic exponent ρ of $A(t)$.*

Proof. Follows from Theorems 2.8 and 2.13 in Chapter 2.   □

The Stability Theorem for Linear Systems reduces the problem of determining whether the solution 0 of equation (3) is stable or asymptotically stable to the problem of studying the real parts of the roots of polynomial equations. This is by no means a simple problem, especially if the polynomial is of high degree, say, for example, degree 67. The problem has been discussed at length by Marden [1966]. Practical topological criteria have been obtained by Cesari [1971] and Cronin [1971]. Here we will just state one important and widely used test.

Routh-Hurwitz Criterion

If

$$P(z) = z^n + a_1 z^{n-1} + \cdots + a_{n-1}z + a_n$$

is a polynomial with real coefficients, let $D_1 \ldots, D_n$ denote the following determinants.

$$D_1 = a_1$$

$$\cdots$$

$$D_k = \begin{vmatrix} a_1 & a_3 & a_5 & \cdots & a_{2k-1} \\ 1 & a_2 & a_4 & \cdots & a_{2k-2} \\ 0 & a_1 & a_3 & \cdots & a_{2k-3} \\ 0 & 1 & a_2 & \cdots & a_{2k-4} \\ & & \cdot & \cdot & \cdot \\ 0 & 0 & 0 & \cdot & a_k \end{vmatrix} \quad (k = 2, 3, \ldots, n)$$

where $a_j = 0$ if $j > n$. If $D_k > 0$, $k = 1, \ldots, n$, then all the solutions of the equation

$$P(z) = 0$$

have negative real parts.

Proof. See Marden [1966, Chapter IX].

Theorem 4.2 *A necessary condition that all the solutions of $P(z) = z^n + a_1 z^{n-1} + \cdots + a_{n-1} z + a_n = 0$, where $a_1, a_2, \ldots, a_n$ are real, have negative real parts is that $a_1, a_2, \ldots, a_n$ are positive.*

Proof. If the solutions all have negative real parts, then

$$P(z) = \prod_{q=1}^{k} (z + \alpha_q) \prod_{j=1}^{m} (z + \beta_j - i\gamma_j)(z + \beta_j + i\gamma_j)$$

$$= \prod_{q=1}^{k} (z + \alpha_q) \prod_{j=1}^{m} (z^2 + 2\beta_j z + \beta_j^2 + \gamma_j^2)$$

where $\alpha_q > 0$ for $q = 1, \ldots, k$, and $\beta_j > 0$ for $j = 1, \ldots, m$. □

In order to obtain a few more results for linear systems and also to obtain in the next section fundamental results for nonlinear systems, we will use:

Gronwall's Lemma. *If u, v are real-valued nonnegative continuous functions with domain $\{t / t \geq t_0\}$ and if there exists a constant $M \geq 0$*

such that for all $t \geq t_0$

$$u(t) \leq M + \int_{t_0}^{t} u(s)v(s)ds \tag{4}$$

then

$$u(t) \leq M \exp \int_{t_0}^{t} v(s)ds \tag{5}$$

Proof. See Exercise 13 of Chapter 1.

Theorem 4.3 *If 0, the identically zero solution of the linear equation*

$$x' = Ax \tag{6}$$

where A is a constant matrix, is asymptotically stable, then there exists $\varepsilon_0 > 0$ such that if the matrix $C(t)$ is continuous for all $t > -\delta$, where $\delta > 0$, and if $|C(t)| < \varepsilon_0$ for all $t > -\delta$, then the equilibrium point $x = 0$ of the equation

$$x' = [A + C(t)]x \tag{7}$$

is asymptotically stable.

Proof. Let $x(t)$ be an arbitrary fixed solution of (7). By the variation of constants formula we may write:

$$x(t) = y(t) + X(t) \int_{0}^{t} [X(s)]^{-1}C(s)x(s)ds \tag{8}$$

where $y(t)$ is the solution of $y' = Ay$ such that $y(0) = x(0)$ and $X(t)$ is the fundamental matrix of $y' = Ay$ such that $X(0) = I$, the identity matrix. Then if s is a positive constant, the matrix

$$X(t)[X(s)]^{-1}$$

is a fundamental matrix and at $t = s$,

$$X(t)[X(s)]^{-1} = I$$

The matrix $X(t - s)$ is also a fundamental matrix such that at $t = s$, the matrix is the identity matrix. Hence by uniqueness of solutions, we can conclude that for all real t,

$$X(t)[X(s)]^{-1} = X(t - s)$$

So (8) can be written:

$$x(t) = y(t) + \int_0^t X(t - s)C(s)x(s)ds \qquad (9)$$

Since the solution of $x' = Ax$ is asymptotically stable then by 2. of the Stability Theorem for Linear Systems the eigenvalues of A all have negative real parts. Hence there exist positive constants a and k such that for all $t \geq 0$,

$$|X(t)| \leq ke^{-at}$$

It follows that if $|y(0)| = k_1$, then for all $t \geq 0$, $|y(t)| \leq k_1 ke^{-at}$. Then from (9), we have:

$$|x(t)| \leq |y(t)| + \int_0^t |X(t - s)|\,|C(s)|\,|x(s)|ds$$

$$\leq k_1 ke^{-at} + \int_0^t \varepsilon_0 ke^{-a(t-s)}|x(s)|ds$$

and

$$|x(t)|e^{at} \leq k_1 k + \int_0^t \varepsilon_0 ke^{as}|x(s)|ds$$

By Gronwall's Lemma,

$$|x(t)|e^{at} \leq k_1 k \exp \int_0^t \varepsilon_0 k ds = k_1 k \exp(\varepsilon_0 k t)$$

or

$$|x(t)| \leq k_1 k \exp[(\varepsilon_0 k - a)t]$$

Then if

$$\varepsilon_0 < \frac{a}{k}$$

the desired result holds. Note that the bound on ε_0 is independent of the solution $x(t)$. This completes the proof of Theorem 4.3. □

Theorem 4.4 *If all the solutions of*

$$x' = Ax$$

where A is a constant matrix, are bounded on the right, if the matrix $C(t)$ is continuous for $t > -\delta$ and if $\int_0^\infty |C(t)| dt < \infty$, then all the solutions of

$$x' = [A + C(t)]x$$

are bounded on the right.

Proof. As in the proof of Theorem 4.3, we have:

$$x(t) = y(t) + \int_0^t X(t - s)C(s)x(s)ds$$

By hypothesis, there exist positive constants c_1 and c_2 such that for all $t \geq 0$,

$$|y(t)| < c_1$$

and

$$|X(t)| < c_2$$

Since

$$|x(t)| \leq |y(t)| + \int_0^t |X(t-s)|\,|C(s)|\,|x(s)|ds$$

or

$$|x(t)| \leq c_1 + \int_0^t c_2|C(s)|\,|x(s)|ds$$

then by Gronwall's Lemma, for all $t \geq 0$,

$$|x(t)| \leq c_1 \exp \int_0^t c_2|C(s)|ds$$

$$\leq c_1 \exp \left\{ c_2 \int_0^\infty |C(s)|ds \right\} \qquad \square$$

Stability of Solutions of Nonlinear Systems

The results we obtain in this chapter for nonlinear systems are essentially obtained by approximating the nonlinear system with a linear system. Thus the Stability Theorem for Linear Systems obtained in the preceding section plays an important role. We consider the n-dimensional equation

$$x' = f(t, x) \tag{10}$$

where $f(t, x)$ has domain D, an open set in (t, x)-space, i.e., R^{n+1}, such that

$$D \supset \{(t, 0) \in R^{n+1} / t \geq 0\}$$

and if f_i denotes the ith component of f, then $\partial f_i / \partial x_j$ $(i, j = 1, \ldots, n)$ exists and is continuous at each point of D. Suppose

$$x(t) = x(t, t_0, x^0)$$

is a solution of (10) which is defined for all $t \geq t_0 \geq 0$. If $\bar{x}(t)$ is another solution of (10) let the function $u(t)$ be defined by

$$\bar{x}(t) = x(t) + u(t)$$

Since $\bar{x}(t)$ is a solution of (10), then

$$x' + u' = f(t, x + u) \tag{11}$$

Since

$$x' = f(t, x) \tag{12}$$

then subtracting (12) from (11), we obtain

$$u' = f(t, x + u) - f(t, x) \tag{13}$$

If we denote the ith component of the right hand side of (13) by $F_i(t, u)$, then

$$F_i(t, u) = \sum_j a_{ij}(t) u_j + X_i(t, u)$$

where

$$a_{ij}(t) = \frac{\partial f_i}{\partial x_j}[t, x(t)]$$

u_j is the jth component of $u(t)$ and $X_i(t, u)$ is continuous in t and u and

$$\lim_{|u| \to 0} \frac{X_i(t, u)}{|u|} = 0$$

for each $t \geq t_0$. Thus (13) becomes:

$$u' = [a_{ij}(t)]u + X(t, u) \tag{14}$$

where $X(t, u)$ has components $X_1(t, u), \ldots, X_n(t, u)$.

Definition. Equation (14) is the *variational system of* (10) *relative to solution* $x(t)$. The linear equation

$$u' = [a_{ij}(t)]u \qquad (15)$$

is the *linear variational system* of (10) relative to $x(t)$.

Notice that $x(t)$ is a stable [asymptotically stable] solution of (10) iff $u(t) \equiv 0$ is a stable [asymptotically stable] solution of (14).

Definition. If $u(t) \equiv 0$ is a stable solution of the linear variational system (15), then solution $x(t)$ of (10) is *infinitesimally stable.*

Infinitesimal stability refers to the stability of the linear approximation of (10) that holds in a neighborhood of the given solution $x(t)$. It is rather reasonable to conjecture that infinitesimal stability implies stability. But this reasonable conjecture is not always valid as the following examples show:

1. The linear variational system of the equation $x' = x^2$ relative to the solution $x(t) \equiv 0$ is $u' = 0$. The solution $u(t) \equiv 0$ of $u' = 0$ is a stable solution, but $u(t) \equiv 0$ is an unstable solution of $x' = x^2$. (See Example 3 earlier in this chapter.)

2. The linear variational system of the equation

$$\begin{aligned} x' &= -y + x^3 + xy^2 \\ y' &= x + y^3 + x^2 y \end{aligned} \qquad (16)$$

   relative to the solution $x(t) \equiv 0$, $y(t) \equiv 0$ is

   $$\begin{aligned} x' &= -y \\ y' &= x \end{aligned}$$

   Since the equilibrium point $(0,0)$ of this system is a center, the solution $x(t) \equiv 0$, $y(t) \equiv 0$ of (16) is infinitesimally stable, but $x(t) \equiv 0$, $y(t) \equiv 0$ is an unstable solution of equation (16) because

   $$rr' = x(-y + x^3 + xy^2) + y(x + y^3 + x^2 y) = (x^2 + y^2)^2 = r^4 > 0$$

   (Since $r' > 0$, all solutions "move away" from $(0,0)$ with increasing time.)

These examples show that infinitesimal stability does not imply stability. But we will show that if $[a_{ij}(t)]$ is a constant matrix or a periodic matrix, then if $u(t) \equiv 0$ is asymptotically stable, the solution $x(t)$ is asymptotically stable; and if $u(t) \equiv 0$ is unstable, then $x(t)$ is unstable. In order to state these results, it is convenient to introduce the following "little o" definition.

Definition. Suppose the vector function $h(t,x)$ is defined on an open set D in (t,x)-space such that

$$D \supset \{(t,0) \mid t \in E\}$$

where D is a subset of the t-axis. Then $h(t,x)$ *satisfies the condition*

$$|h(t,x)| = o(|x|)$$

as $|x| \to 0$ uniformly with respect to $t \in E$ means: there exists a real-valued function $m(r)$ defined for $r > 0$ such that:

(1) $m(r) > 0$ for all $r > 0$;

(2) $\lim\limits_{r \to 0} m(r) = 0$;

and there exists $\bar{\varepsilon} > 0$ such that if $|x| < \bar{\varepsilon}$, then for all $t \subset E$,

$$|h(t,x)| \leq [m(|x|)]|x|$$

Asymptotic Stability Theorem for Nonlinear Systems. *Suppose that the eigenvalues of matrix A are $\lambda_1, \ldots, \lambda_m$ and that $R(\lambda_i) < 0$ for $i = 1, \ldots, m$. Suppose also that $h(t,x)$ is continuous on the set*

$$[(t,x)/t > -\delta, \ x \in U]$$

where $\delta > 0$ and U is a neighborhood in R^n of $x = 0$ and $h(t,x)$ satisfies a Lipschitz condition in x, and

$$|h(t,x)| = o(|x|)$$

as $|x| \to 0$ *uniformly with respect to t where* $t > -\delta$. *Then the solution* $x(t) \equiv 0$ *of the equation*

$$x' = Ax + h(t,x) \tag{17}$$

is asymptotically stable.

Proof. Let $a < 0$ be a negative number such that

$$R(\lambda_i) < a \qquad (i = 1, \ldots, m)$$

and let $Y(t)$ be the fundamental matrix of $y' = Ay$ such that $Y(0) = I$, the identity matrix. Then there is a positive number C such that for all $t \geq 0$,

$$|Y(t)| \leq Ce^{at}$$

Suppose a solution $x(t)$ of (17) is defined in an open interval I containing $t = 0$. By the variation of constants formula, if $t \in I$,

$$x(t) = y(t) + \int_0^t Y(t-s)h[s,x(s)]ds$$

But

$$|y(t)| = |Y(t)y(0)| = |Y(t)x(0)| \leq Ce^{at}|x(0)|$$

and hence

$$|x(t)| \leq Ce^{at}|x(0)| + \int_0^t |Y(t-s)|\,|h[s,x(s)]|ds$$

Now given a constant $M > 0$, then there exists $d = d(M) > 0$ such that if $t \in I$ and $|x| \leq d$, then

$$|h(t,x)| \leq M|x|$$

If $|x(0)| < d$, then there exists $\delta = \delta(d)$ such that if $t \in (-\delta, \delta)$, then

$$|x(t)| < d$$

Hence if $t \in (-\delta, \delta)$, then

$$|x(t)| \leq C|x(0)|e^{at} + \int_0^t Ce^{a(t-s)}M|x(s)|ds$$

Since $a = -b$ where $b > 0$, we may write:

$$|x(t)|e^{bt} \leq C|x(0)| + \int_0^t MCe^{bs}|x(s)|ds$$

and hence by Gronwall's Lemma, if $0 \leq t < \delta$,

$$|x(t)|e^{bt} \leq C|x(0)|\exp\int_0^t MCds$$

or

$$|x(t)| \leq C|x(0)|\exp[(MC - b)t]$$

and if

$$M < \frac{b}{C}$$

then

$$|x(t)| \leq C|x(0)|$$

Thus we have proved the statement: if M is a positive constant such that $M < b/C$, then there exists $d = d(M) > 0$ and $\delta > 0$ such that if $|x(0)| < d$ and $t \in [0, \delta)$, then $|x(t)| < d$ and

$$|x(t)| < C|x(0)|$$

From this statement we obtain at once the following lemma.

Lemma 4.1 *Let $d = d(M)$ and let*

$$r = \min\left(d, \frac{d}{2C}\right)$$

Let $|x(0)| < r$ and let $\delta_1 > 0$ be such that if $t \in [0, \delta_1)$ then

$$|x(t)| < d$$

Then if $t \in [0, \delta_1)$,

$$|x(t)| < C|x(0)| < Cr \leq \frac{d}{2}$$

Now we prove that $x(t) \equiv 0$ is asymptotically stable. To verify conditions (1) and (2), consider solution $x(t)$ such that $|x(0)| < r$, where $r = \min\left(d, \frac{d}{2C}\right)$ and $d \leq d(M)$ and $m < \frac{b}{C}$. Let

$$t_0 = \text{lub}\Big\{t_1/t_1 \geq 0 \text{ and if } t \in [0, t_1) \text{ then solution } x(t)$$

$$\text{is defined and } |x(t)| < d\Big\}$$

Then $t_0 > 0$ because $|x(0)| < d$ and $x(t)$ is continuous. Now suppose t_0 is finite. By Lemma 4.1, if $t \in (0, t_0)$, then

$$|x(t)| < \frac{d}{2}$$

It follows that the solution $x(t)$ can be extended so that its domain I contains t_0 and, by the continuity of the solution,

$$|x(t)| < d$$

for all $t \in I$. This contradicts the properties of t_0. Thus conditions (1) and (2) in the definition of stability are established. To verify condition (3), notice that we have just shown that if $|x(0)| < r$, then $|x(t)| < d$ for all $t \geq 0$. We proved earlier that if $|x(t)| < d$, then

$$|x(t)| < C|x(0)| \exp[(MC - b)t]$$

Hence if $|x(0)| < r$, then for all $t \geq 0$,

$$|x(t)| < C|x(0)| \exp[(MC - b)t]$$

Since $MC - b < 0$,

$$\lim_{t \to \infty} |x(t)| = 0$$

Thus condition (3) in the definition of asymptotic stability is satisfied. This completes the proof of the Asymptotic Stability Theorem for Nonlinear Systems. $\square$

Instability Theorem for Nonlinear Systems. *If matrix A has at least one eigenvalue λ such that $\mathcal{R}(\lambda) > 0$, the solution $x(t) \equiv 0$ of*

$$x' = Ax + h(t, x)$$

is unstable.

Proof. We postpone the proof of the theorem until we can use a Lyapunov function. (See Exercise 6 in Chapter 5.)

A corollary of the Stability and Instability Theorems in which the constant matrix A is replaced with a periodic matrix $A(t)$ is easily obtained by using the theory of characteristic multipliers and exponents developed in Chapter 2. We obtain:

Corollary to Stability and Instability Theorems for Nonlinear Systems. *Suppose the matrix $A(t)$ is continuous in t for all real t and has period τ, and suppose that the hypotheses on $h(t, x)$ are the same as in the Stability and Instability Theorems. Let $\lambda_1, \ldots, \lambda_m$ be the characteristic multipliers and $\rho_1, \ldots, \rho_m$ the corresponding characteristic exponents of $A(t)$. If $|\lambda_1| < 1$ ($R(\rho_i) < 0$) for $i = 1, \ldots, m$, the solution $x(t) \equiv 0$ of the equation*

$$x' = A(t)x + h(t, x)$$

is asymptotically stable. If there exists λ_j such that $|\lambda_j| > 1$ ($R(\rho_j) > 0$) then $x(t) \equiv 0$ is not stable.

Proof. Let

$$v = Zx$$

where

$$Z(t) = e^{tR}[X(t)]^{-1}$$

Matrix $Z(t)$ is defined in the proof of Theorem 2.13 and we obtain:

$$\frac{dv}{dt} = Rv + Zh[t, Z^{-1}v] \qquad (18)$$

where the characteristic exponents are the eigenvalues of matrix R. From the properties of $Z(t)$, it follows easily that the Stability and Instability Theorems are applicable to the solution $v(t) \equiv 0$ of (18). If all the eigenvalues of R (all the characteristic exponents) have negative real parts, then $v(t) \equiv 0$ is asymptotically stable and, again from the properties of $Z(t)$, it follows that

$$x(t) = Z^{-1}(t)v(t) \equiv 0$$

is asymptotically stable. If there is an eigenvalue of R with positive real part, a similar argument shows that $x(t) \equiv 0$ is not stable. This completes the proof of the Corollary.   □

The preceding Corollary makes possible the study of stability properties of periodic solutions of nonlinear systems. Suppose we consider the system

$$x' = f(t, x) \qquad (19)$$

where f has continuous first derivatives in an open set D in (t, x)-space and f has period τ in t. Suppose that $\bar{x}(t)$ is a solution of (19) which has period τ and is such that

$$D \supset \{(t, \bar{x}(0))/0 \leq t \leq \tau\}$$

The variational system of (19) relative to $\bar{x}(t)$ is:

$$u' = \left[\frac{\partial f_i}{\partial x_j}[t, \bar{x}(t)]\right] u + X(t, u) \qquad (20)$$

where f_i is the ith component of f and $X(t, u)$ has the same meaning as in equation (14). Since $\bar{x}(t)$ has period τ and $f(t, x)$ has period τ in t, it follows that the matrix

$$M(t) = \left[\frac{\partial f_i}{\partial x_j}[t, \bar{x}(t)]\right]$$

has period τ.

Stability Theorem for Periodic Solutions. *Let* $\lambda_1,\ldots,\lambda_m$ *be the characteristic multipliers and* $\rho_1,\ldots,\rho_m$ *the corresponding characteristic exponents of* $M(t)$. *If* $|\lambda_i| < 1$ $(R(\rho_i) < 0)$ *for* $i = 1,\ldots,m$, *then* $\bar{x}(t)$ *is asymptotically stable. If there exists* λ_j *such that* $|\lambda_j| > 1$ $(R(\rho_j) > 0)$ *rthen* $\bar{x}(t)$ *is not stable.*

Proof. If $|\lambda_i| < 1$ for $i = 1,\ldots,m$, then by the Corollary, $u(t) \equiv 0$ is an asymptotically stable solution of the variational system (20). Hence $\bar{x}(t)$ is asymptotically stable. A similar proof yields the second statement of the theorem. $\square$

Some Stability Theory for Autonomous Nonlinear Systems

There are some unexpected complications when the stability theory developed thus far is applied to solutions of autonomous systems. First, it turns out that there is no possibility of applying the Stability Theorem for Periodic Solutions obtained above. For if the autonomous system

$$x' = f(x) \tag{21}$$

has a solution $\bar{x}(t)$ of period τ, then substituting $\bar{x}(t)$ into (21) and differentiating, we obtain:

$$\frac{d}{dt}\left[\frac{d}{dt}\,\bar{x}_i\right] = \sum_{j=1}^{n} \frac{\partial f_i}{\partial x_j}[\bar{x}(t)]\frac{d\bar{x}_j}{dt} \quad (i = 1,\ldots,n) \tag{22}$$

where $\bar{x}_i$, f_i are the ith components of $\bar{x}$ and f, respectively. Equation (22) shows that $d\bar{x}/dt$ is a solution of the linear variational system of (21) relative to solution $\bar{x}(t)$. Since $\bar{x}(t)$ has period τ, then $d\bar{x}/dt$ has period τ. Hence by Theorem 2.14, at least one characteristic multiplier of the coefficient matrix

$$\left[\frac{\partial f_i}{\partial x_j}[\bar{x}(t)]\right]$$

of (22) has the value 1. Hence the hypothesis for asymptotic stability in the Stability Theorem for Periodic Solutions cannot be satisfied.

Actually we can go considerably further and show that there are no nontrivial asymptotically stable periodic solutions of an autonomous system. That is, we prove the following theorem.

Theorem 4.5 *If $x(t)$ is a nontrivial periodic solution of an autonomous system*

$$x' = f(x) \tag{23}$$

then $x(t)$ is not asymptotically stable.

Proof. Since $x(t)$ is a nontrivial periodic solution, then there exists t_0 such that $f[x(t_0)] \neq 0$. Then

$$\frac{dx}{dt}(t_0) \neq 0 \tag{24}$$

Suppose $x(t)$ is asymptotically stable. Then given $\varepsilon > 0$, let $\bar{\delta}(\varepsilon)$ denote the minimum of the δ in condition (2) of the definition of asymptotic stability and the $\bar{\delta}$ in condition (3) of the definition of asymptotic stability. If $|\bar{t}|$ is sufficiently small and fixed, then

$$|x(t_0) - x(t_0 + \bar{t})| < \bar{\delta}(\varepsilon) \tag{25}$$

Also by (24) there exists $r > 0$ such that

$$|x(t_0) - x(t_0 + \bar{t})| > r \tag{26}$$

By the asymptotic stability of $x(t)$ and the fact that $x(t + \bar{t})$ is a solution of (23) (by Lemma 3.1 of Chapter 3), it follows that

$$\lim_{t \to \infty} |x(t) - x(t + \bar{t})| = 0 \tag{27}$$

but since $x(t)$ has period $\tau > 0$, then by (25) for all integers n,

$$|x(t_0 + n\tau) - x(t_0 + n\tau + \bar{t})| > r$$

and hence (27) cannot be satisfied.   □

Theorem 4.5 shows that we must seek a weaker asymptotic stability condition for use with autonomous systems. The following definitions are sometimes useful.

Definition. Let $x(t)$ be a solution of the system

$$x' = f(x) \tag{28}$$

where f is defined and satisfies a Lipschitz condition in an open set $D \subset R^n$, such that $x(t)$ has period τ. Let C be the orbit of $x(t)$ and if $p \in R^n$, let

$$d(p,C) = \mathrm{g\ell b}_{q \in C} \, d(p,q)$$

where $d(p,q)$ is the usual Euclidean distance between the points p and q in R^n. Then the orbit C is *orbitally stable* if: $\varepsilon > 0$ implies that there exists $\delta > 0$ such that for any solution $x^{(1)}(t)$ of (28) for which there exists t_0 such that

$$d(x^{(1)}(t_0), C) < \delta$$

it is true that $x^{(1)}(t)$ is defined for all $t \geq t_0$ and

$$d(x^{(1)}(t), C) < \varepsilon$$

for all t. Orbit C is *asymptotically orbitally stable* if C is orbitally stable and there exists $\varepsilon_0 > 0$ such that for any solution $x(t)$ for which there exists t_0 such that

$$d(x(t_0), C) < \varepsilon_0$$

it is true that

$$\lim_{t \to \infty} d(x(t), C) = 0$$

Orbital stability is a considerably weaker condition than stability. Roughly speaking, it says that solutions stay close together in a point set sense but the positions of the solutions for equal values of t may get very far apart. (For an example of a periodic solution which is orbitally stable but not stable, see Hahn [1967, p. 172].) The notion of orbital stability is useful only if the orbit C is a closed curve, i.e., C is described by a periodic solution. Otherwise the orbit C may "cover so much territory" that orbital stability is meaningless.

A stronger stability condition which can be imposed on solutions of autonomous systems and which is somewhat closer to an intuitively desirable stability condition is the following:

Definition. Let $x(t)$ be a solution of the system

$$x' = f(x) \tag{29}$$

where f is defined and continuous on an open set $D \subset R^n$. Then $x(t)$ is *uniformly stable* if there exists a constant K such that $\varepsilon > 0$ implies there is a positive $\delta(\varepsilon)$ so that if $u(t)$ is a solution of (29) and if there exist numbers t_1, t_2 such that $t_2 \geq K$ and such that

$$|u(t_1) - x(t_2)| < \delta(\varepsilon)$$

then for all $t \geq 0$

$$|u(t + t_1) - x(t + t_2)| < \varepsilon$$

Solution $x(t)$ of (29) is *phase asymptotically stable* if there exists a cosntant K such that $\varepsilon > 0$ implies there is a positive $\delta(\varepsilon)$ so that if $u(t)$ is a solution of (29) and if there exist numbers t_1, t_2 such that $t_2 \geq K$ and such that

$$|u(t_1) - x(t_2)| < \delta(\varepsilon)$$

then for all $t \geq 0$,

$$|u(t + t_1) - x(t + t_2)| < \varepsilon$$

and there exists a number t_3 such that

$$\lim_{t \to \infty} |u(t) - x(t_3 + t)| = 0 \tag{P}$$

Roughly speaking, solution $x(t)$ is phase asymptotically stable if $x(t)$ is uniformly stable and condition (P) holds.

Phase Asymptotic Stability Theorem for Periodic Solutions. *Suppose*

$$x' = g(x) \tag{30}$$

is such that g has continuous third derivatives in an open set D in R^n and suppose that $\bar{x}(t)$ is a solution of (30) such that $\bar{x}(t)$ has period

$\tau > 0$ *and such that the linear variational system of (30) relative to* $\bar{x}(t)$ *has* $(n-1)$ *characteristic exponents* $\rho_1, \ldots, \rho_{n-1}$ *such that*

$$R(\rho_j) < 0 \qquad (j = 1, \ldots, n-1)$$

Then $\bar{x}(t)$ *is phase asymptotically stable.*

Proof. We note first that, since (30) is autonomous, then as pointed out at the beginning of this section, at least one characteristic multiplier of the linear variational system equals one. Hence at least one characteristic exponent is zero and the hypothesis of the theorem could be stated as: zero is a characteristic exponent of algebraic multiplicity one and all the other characteristic exponents have negative real parts.

By translation of coordinate axes, we may assume that $\bar{x}(0) = 0$ and by a rotation of coordinate axes, we may assume that $\bar{x}_1'(0) \neq 0$, $\bar{x}_j'(0) = 0$ for $j = 2, \ldots, n$. By a further change of coordinates using Floquet theory we may write the variational equation of (30) relative to $\bar{x}(t)$ as:

$$u' = Ru + f(t, u) \tag{31}$$

where

$$R = \begin{bmatrix} 0 & 0 \\ 0 & B \end{bmatrix}$$

where the eigenvalues of B are the characteristic exponents which have negative real parts and

$$\lim_{|u - u^{(1)}| \to 0} \frac{|f(t, u) - f(t, u^{(1)})|}{|u - u^{(1)}|} = 0$$

uniformly in t.

In order to prove the theorem, we obtain first a preliminary result. Suppose that vector a has the form $(0, a_2, \ldots, a_n)$ and consider the integral equation

$$w(t, a) = e^{tR}a + \int_0^t \begin{bmatrix} 0 & 0 \\ 0 & e^{(t-s)B} \end{bmatrix} f[s, w(s, a)]ds$$

$$- \int_t^\infty \begin{bmatrix} 1 & 0 \\ 0 & 0 \end{bmatrix} f[s, w(s, a)]ds$$

(32)

We show that if $|a|$ is sufficiently small and $t \geq 0$, then (32) can be solved for $w(t, a)$ and

$$\lim_{t \to \infty} w(t, a) = 0$$

uniformly in a. The proof is a standard use of successive approximations. We define the sequence

$$w_0(t, a) = 0$$

$$w_{k+1}(t, a) = e^{tR}a + \int_0^t \begin{bmatrix} 0 & 0 \\ 0 & e^{(t-s)B} \end{bmatrix} f[s, w_k(s, a)]ds$$

$$- \int_t^\infty \begin{bmatrix} 1 & 0 \\ 0 & 0 \end{bmatrix} f[s, w_k(s, a)]ds$$

Since $a = (0, a_2, \ldots, a_n)$, there exists $M > 1$ such that

$$|e^{tR}a| \leq M|a|e^{-\sigma t}$$

where $\sigma > 0$ and $R(\lambda) < -\sigma$ for all eigenvalues λ of B. Then it follows that

$$|w_1(t, a) - w_0(t, a)| \leq M|a|e^{-\sigma t}$$

Since f is a "higher order" term, uniformly in t, there exists $\delta > 0$ such that if $|u| < \delta$, $|u^{(1)}| < \delta$, then

$$|f(t, u) - f(t, u^{(1)})| < \frac{\sigma}{8M}|u - u^{(1)}|$$

It is a straightforward computation to prove by induction (see Exercise 8) that if $t \geq 0$ and $|a|$ is sufficiently small, $|a| < 1$, then for $k = 1, 2, \ldots$,

$$|w_{k+1}(t, a) - w_k(t, a)| \leq \frac{M|a|e^{-\frac{\sigma}{2}t}}{2^k}$$

It follows that the sequence $\{w_k(t,a)\}$ converges uniformly in t to a function $w(t,a)$ and

$$|w(t,a)| \leq 2M|a|e^{-\frac{\varepsilon}{2}t}$$

Thus $w(t,a)$ is a solution of (32) and

$$\lim_{t \to \infty} w(t,a) = 0$$

uniformly in a for $|a| < 1$. Straightforward computation shows that $w(t,a)$ is a solution of (31). From (32), we have:

$$w(0,a) = a - \int_0^\infty \begin{bmatrix} 1 & 0 \\ 0 & 0 \end{bmatrix} f[s, w(s,a)]ds \tag{33}$$

Denoting $w(0,a)$ by $(v_1, \ldots, v_n)$, we may rewrite (33) as

$$v_1 = -\int_0^\infty f_1[s, w(s, a_2, \ldots, a_n)]ds$$

$$v_2 = a_2$$

$$v_n = a_n$$

Then we may restate the results concerning $w(t,a)$ as

Lemma 4.2 *Let $w(t,a)$ where $a = (0, a_2, \ldots, a_n)$ be the solution (obtained above) of (32) such that*

$$w(0,a) = (a_1, a_2, \ldots, a_n)$$

where

$$a_1 = -\int_0^a f_1[s, w(s, a_2, \ldots, a_n)]ds \tag{34}$$

then $w(t,a)$ is such that

$$\lim_{t \to \infty} w(t,a) = 0$$

uniformly for $|a| < 1$. Also $w(t,a)$ is a solution of (31).

Now we are ready to prove the theorem. Let Σ denote the surface in $(a_1, \ldots, a_n)$-space described by (34). Suppose $x(t)$ is a solution such that for some t_1,

$$|x(t_1) - \bar{x}(0)| = |x(t_1)| < \varepsilon$$

By reparameterizing $x(t)$, we may choose $t_1 = 0$. We prove first that if ε is sufficiently small, then $x(t)$ intersects the surface Σ. That is, we show that if ε is sufficiently small, then the equation

$$x_1(t) + \int_0^\infty f_1[s, w(s, a_2, \ldots, a_n)]ds = 0 \qquad (35)$$

can be solved for t as a function of $a_2, \ldots, a_n$. Equation (35) has the initial solution

$$t = 0, \; a_2 = \cdots = a_n = 0$$

because $w(s, 0) = 0$. If ε is sufficiently small, then

$$x_1'(0) \neq 0.$$

Hence by the Implicit Function Theorem, equation (35) can be solved locally for t as a function of $(a_2, \ldots, a_n)$. Thus we have

$$x[t(a_2, \ldots, a_n)] = \left(-\int_0^\infty f_1[s, w(s, a_2, \ldots, a_n)]ds, a_2, \ldots, a_n \right)$$

Now take a fixed $x(t)$. Reparameterize so that

$$x(0) = \left(-\int_0^\infty f_1[s, w(s, a_2 \ldots, a_n)]ds, a_2, \ldots, a_n \right)$$

Let

$$u(t) = x(t) - \bar{x}(t)$$

Then

$$u(0) = \left(-\int_0^\infty f_1[s, w(s, a_2, \ldots, a_n)]ds, a_2, \ldots, a_n \right) \qquad (36)$$

Since $u(t)$ is a solution of (31), then (36) shows that

$$u(t) = w(t, a)$$

or

$$x(t) - \bar{x}(t) = w(t, a)$$

The proof of the theorem follows from the properties of $w(t, a)$. $\square$

Another serious limitation of the Lyapunov stability is that the notions of asymptotic stability, phase asymptotic stability and asymptotic orbital stability cannot be applied to Hamiltonian systems, i.e., a solution of a Hamiltonian system is neither asymptotically stable, phase asymptotically stable nor asymptotically orbitally stable. More precisely, we have the following:

Definition. A *time-independent Hamiltonian system* is a $(2n)$-dimensional system

$$\begin{aligned} x' &= H_y(x, y) \\ y' &= -H_x(x, y) \end{aligned} \qquad (37)$$

where x, y are n-vectors, $H(x, y)$ is a real-valued function of x, y which has continuous first derivatives with respect to each of the components of x and y, $H_x(x, y)$ denotes the n-vector $(\partial H/\partial x_1, \ldots, \partial H/\partial x_n)$ where $x_1, \ldots, x_n$ are the components of x, and $H_y(x, y)$ has a similar meaning. Thus as a system of scalar equations, the

Hamiltonian system is:

$$x_1' = \frac{\partial H}{\partial y_1} \qquad\qquad y_1' = \frac{-\partial H}{\partial x_1}$$

$$\cdots \qquad\qquad\qquad \cdots$$

$$x_n' = \frac{\partial H}{\partial y_n} \qquad\qquad y_n' = -\frac{\partial H}{\partial x_n}$$

Theorem 4.6 *If* $(x(t), y(t))$ *is a solution of a time-independent Hamiltonian system, then* $(x(t), y(t))$ *is neither asymptotically stable, phase asymptotically stable nor asymptotically orbitally stable.*

Proof. Suppose that $(x(t), y(t))$ is a solution of (37) which is asymptotically stable. Then there exists a number t_0 and a positive number δ such that if $(\bar{x}(t), \bar{y}(t))$ is a solution of (37) and if

$$|\bar{x}(t_0) - x(t_0)| + |\bar{y}(t_0) - y(t_0)| < \delta$$

then

$$\lim_{t \to \infty} [|\bar{x}(t) - x(t)| + |\bar{y}(t) - y(t)|] = 0 \tag{38}$$

But if $x(t), y(t))$ is a solution of (37), then $H[x(t), y(t)]$ is a constant because

$$\frac{d}{dt}\{H[x(t), y(t)]\} = \frac{\partial H}{\partial x}\frac{dx}{dt} + \frac{\partial H}{\partial y}\frac{dy}{dt}$$

$$= H_x H_y + H_y(-H_x)$$

$$= 0$$

By the continuity of H, it follows from (38) that $H(x, y)$ is constant in a neighborhood N_δ of radius δ of the point $(x(t_0), y(t_0))$. Thus every point in N_δ is an equilibrium point of (37). The solution $(x(t), y(t))$ is the equilibrium point $(x(t_0), y(t_0))$ and is clearly not asymptotically stable because each point in N_δ is an equilibrium point. A similar contradiction is obtained if we assume that $(x(t), y(t))$ is phase asymptotically stable or asymptotically orbitally stable. This completes the proof of Theorem 4.6 $\qquad \square$

Asymptotic stability of the kind discussed here seems an intuitively reasonable condition to impose. However, it should be pointed out that while the stability concepts introduced in this chapter have the advantage that they can be used as the basis for a coherent and attractive mathematical theory, they are not satisfactory from the physical viewpoint. Lyapunov stability is neither necessary nor sufficient for stability in a physical sense. For example, suppose that a physical system is described by an equation

$$x' = f(t, x) \tag{39}$$

and $x(t)$ is a stable solution of the equation. Suppose that, for some reasonable value of ε, say $\varepsilon = 1/2$, it turns out that the corresponding $\delta(\varepsilon)$ is 10^{-25}. From the mathematical viewpoint, this is not important, but from the physical viewpoint, the solution $x(t)$ can hardly be regarded as stable. On the other hand, the solution $x(t)$ may possess useful stability properties even if it is not stable in the rigorous mathematical sense. For example, if equation (39) is a 2-dimensional autonomous system and $x(t)$ is an equilibrium point p and the orbits around p are as sketched in Figure 1, then if radius R_2 is very small and radius R_1 is very large, the equilibrium point p has useful stability properties even though it is not Lyapunov stable. (Solutions leave the circle with small radius R_2, but no solution escapes the circle with radius R_1. Moreover, solution in the large circle move toward the smaller circle.) The equilibrium point p is said to be practically stable. For a discussion of practical stability, see Hahn [1967].

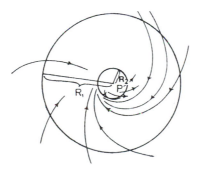

Figure 1

Throughout this chapter, we have taken the view that the solutions of the differential equation which have physical significance are those which have stability properties. Another viewpoint is to look at the differential equation itself. Instead of imagining a disturbance such that the physical system being described is moved from one solution onto another solution, we consider that the differential equation itself is perturbed, i.e., the original differential equation

$$x' = f(x)$$

is replaced by the equation

$$x' = f(x) + p(x)$$

where $p(x)$ represents a perturbation and we ask if there exists a homeomorphism from the orbits of the equation

$$x' = f(x)$$

to the orbits of the equation

$$x' = f(x) + p(x)$$

If for "small" perturbations $p(x)$, there exists such a homeomorphism, the equation

$$x' = f(x)$$

is said to be *structurally stable*. (For a rigorous and enlightening discussion of structural stability, see Hirsch and Smale [1974], Chapter 16.) Then instead of seeking stable solutions, we would search for a structurally stable differential equation to describe the physical system.

However, it should be emphasized that neither the viewpoint of studying stable solutions nor the viewpoint of studying structurally stable differential equations is always satisfactory. Certainly for some purposes, the global property of structural stability is better, both for applications and for aesthetic reasons. On the other hand, there are problems important in applications in which the differential equation that must be studied is not structurally stable. This occurs, for example, in bifurcation problems which will be considered in detail in Chapter 7.

Exercises for Chapter 4

1. Sketch some of the orbits near the origin of solutions of the following systems:

   a) $x' = 2x$

   $y' = -3y$

   $z' = -2z$

   b) $x' = 2x + y$

   $y' = -x + 2y$

   $z' = 3z$

   c) $x' = 2x + y$

   $y' = 2y + z$

   $z' = 2z$

2. Prove the statements in Remark 2 following the definitions of stability and asymptotic stability. (Hint: impose enough hypotheses so that the solution $x(t, t_0, x^0)$ is unique and is continuous in x^0. Then prove that if A is a real number and $B_r(x^0)$ is closed ball in R^n with center x^0 and radius r such that the mapping

$$M : \bar{x} \to x(t_0 + A, t_0, \bar{x})$$

is defined on B_r, then M is a homeomorphism on B_r. The continuity of M follows from the continuity of $x(t, t_0, x^0)$ in x^0. The fact that M is $1-1$ follows from the uniqueness of solution $x(t, t_0, x^0)$.)

3. Show that Theorem 4.3 is not true if the term "asymptotically stable" is replaced with "stable" (in the hypothesis and the conclusion) by showing that two linearly independent solutions of the equation

$$x'' - \frac{2}{t + M} x' + x = 0 \qquad (t \geq 0)$$

where M is a positive constant, are

$$\sin t - (t + M)\cos t$$

and

$$\cos t + (t + M)\sin t$$

4. Given the linear homogeneous system

$$x' = A(t)x \tag{E}$$

where $A(t)$ is continuous for all real t, prove that if there exists a stable solution $x(t)$ of (E), then all solutions of (E) are bounded for $t \geq t_0$, a fixed value. Prove that if (E) has a fundamental matrix which is bounded for $t \geq t_0$, then each solution of (E) is stable. Give an example of an equation (not necessarily linear homogeneous) for which each solution is stable, but no solution is bounded for $t \geq t_0$.

5. Investigate the stability of the equilibrium points of the Volterra equations. (See Exercise 5 at the end of Chapter 3.) That is, determine conditions on the coefficients a, A, c, d, e so that the equilibrium points are asymptotically stable or unstable.

6. Investigate the stability of the equilibrium point $(0,0,0)$ of the Field-Noyes equations.

7. Investigate the stability of the equilibrium point $(0,0,0)$ of the Goodwin equations if $n = 3$, $\rho = 1$, $\alpha = 1$.

8. Prove the inequality

$$|w_{k+1}(t,a) - w_k(t,a)| \leq \frac{M|a|e^{-\frac{s}{2}t}}{2^k}$$

which appears in the proof of the Phase Asymptotic Stability Theorem for Periodic Solutions.

Chapter 5

The Lyapunov Second Method

Definition of Lyapunov Function

We turn now to an entirely different method for studying the stability properties of a given solution. This method, the Lyapunov second method or direct method, uses an approach different from that used in the preceding chapter. It requires no study of the linear parts of equations (eigenvalues or characteristic exponents) and no knowledge is needed of solutions near the solution under study. Of course another price has to be paid: application of Lyapunov's second method demands the construction of the Lyapunov functions to be described later. Such constructions are generally difficult and have been made only for certain classes of equations.

Lyapunov's studies stemmed from consideration of astronomical problems. but the second method has proved very valuable in applications to other problems, e.g., control theory. The basic idea of the second method is to generalize the statement that if the potential energy of a physical system is a minimum [maximum] at an equilibrium point, then the equilibrium point is stable [unstable]. A precise version of this statement is called Lagrange's Theorem. For a discussion of the relation between the concept of stability and Lagrange's Theorem and for a proof of Lagrange's Theorem, see Lefschetz [1962].

In the Lyapunov second method, the potential energy function is replaced by a more general kind of function, the Lyapunov function. Our object here is simply to present detailed proofs of the basic theorems of the second method and indicate some applications. For a beautiful and suggestive account of the theory and some applications, especially to control theory, see LaSalle and Lefschetz [1961]. For an extensive treatment of applications, see Aggarwal and Vidyasagar [1977].

We consider an n-dimensional system

$$x' = f(t, x) \tag{1}$$

where f is defined and satisfies a local Lipschitz condition with respect to x at each point of the set

$$S = \{(t, x)/t_1 < t, |x| < a\}$$

where a is a positive constant and t_1 is a constant. We assume that $x(t) \equiv 0$ is a solution of (1), that is, we assume that for all $t > t_1$,

$$f(t, 0) = 0$$

We will describe how the Lyapunov second method can be used to study the stability properties of the solution $x(t) \equiv 0$ of (1). We need first some definitions. Let the real-valued function

$$V(t, x_1, \ldots, x_n) = V(t, x)$$

have the domain

$$S_1 = \{(t, x)/t_2 < t, |x| < A\}$$

where $t_2 \geq t_1$ and $0 < A < a$ and assume that:

(1) $V(t, x)$ has continuous first partial derivatives with respect to $t, x_1, \ldots, x_n$ at every point of S_1;

(2) $V(t, 0, \ldots, 0) = 0$ for $t > t_2$.

Definitions. The function V is *positive [negative] semidefinite* on S_1 if for all $(t, x) \in S_1$,

$$V(t, x) \geq 0 \ [\leq 0]$$

The function $W(x) = W(x_1, \ldots, x_n)$, satisfying the same hypotheses as $V(t, x)$ but assumed to be independent of t, is *positive [negative] definite on S_1* if $W(x) > 0 \ [< 0]$ for all x such that

$$0 < |x| < A$$

and if $W(0) = 0$. The function $V(t, x)$ is *positive [negative] definite on S_1* if there exists a positive definite function $W(x)$ such that for all $(t, x) \in S_1$,

$$V(t, x) \geq W(x) \ \ [-V(t, x) \geq W(x)]$$

The function $V(t, x)$ is *bounded* if there exists a constant $M > 0$ such that for all $(t, x) \in S_1$,

$$|V(t, x)| < M$$

The function $V(t, x)$ has an *infinitesimal upper bound* if $V(t, x)$ is bounded and $\varepsilon > 0$ implies there exists $\delta > 0$ such that if $|x| < \delta$ and $t > t_2$, then $|V(t, x)| < \varepsilon$.

Remark. The term "infinitesimal upper bound" is misleading and confusing. The property thus termed is simply the property that

$$\lim_{|x| \to 0} V(t, x) = 0$$

uniformly in t for $t > t_2$. However, since the phrase "infinitesimal upper bound" sometimes occurs in the literature, it is necessary to be acquainted with it.

Definition. If $V(t, x)$ has domain S, the function

$$\dot{V} = \dot{V}(t, x_1, \ldots, x_n) = \dot{V}(t, x)$$

is defined to be:

$$\dot{V}(t, x) = \sum_{i=1}^{n} \frac{\partial V}{\partial x_i} f_i(t, x_1, \ldots, x_n) + \frac{\partial V}{\partial t} = \nabla V \cdot f + V_t$$

where $f_i(t, x_1, \ldots, x_n)$ is the i-th component of the function $f(t, x)$ in equation (1).

Remark. If $x(t)$ is a solution of (1), then

$$\dot{V}(t, x(t)) = \sum_{i=1}^{n} \frac{\partial V}{\partial x_i} x_1' + \frac{\partial V}{\partial t}$$

that is, $\dot{V}(t, x(t))$ is the derivative of $V(t, x)$ along the solution $x(t)$.

Theorems of the Lyapunov Second Method

If the function $V(t, x)$ satisfies a definiteness condition and the function $\dot{V}(t, x)$ satisfies a definiteness condition of the opposite sign, the function $V(t, x)$ is called a *Lyapunov function.* Our next step is to show how the existence of Lyapunov functions implies that stability conditions hold.

Stability Theorem. *If there exists a function $V(t, x)$ which is positive definite on S and if $\dot{V}$ is negative semidefinite on S, then $x(t) \equiv 0$ is a stable solution of (1).*

Remark. The geometric meaning of this theorem is clearly seen if we think of a simple case in which $V(t, x)$ is independent of t and is a positive definite quadratic form $Q(x)$ in the components of X. Then if K is a positive constant the equation

$$Q(x) = K$$

is the equation of an ellipsoid. If K is increased the axes of the ellipsoid are increased. But $\dot{V}(t, x)$ is the rate of change of $Q(x)$ along a solution of (1). Hence, since $\dot{V}(t, x)$ is negative semidefinite, then if a solution enters an ellipsoid

$$Q(x) = K$$

it can never "escape" that ellipsoid. Hence, $x(t) \equiv 0$ is stable. The proof of the theorem essentially consists in describing these geometric ideas analytically and rigorously.

Proof of Stability Theorem. **By hypothesis there exists a function** $W(x)$ such that if $t > t_1$ and $0 < |x| < a$, then

$$V(t, x) \geq W(x) > 0$$

and

$$\dot{V}(t, x) \leq 0$$

Let $\bar{t} > t_1$ and let $x(t, \bar{t}, b)$ be the solution of (1) such that $x(\bar{t}, \bar{t}, b) = b$ where $|b| < a$. Then if solution $x(t, \bar{t}, b)$ is maximal, its domain contains $[\bar{t}, t_2]$ where $t_2 = \infty$ or if $t_2 < \infty$ and t_2 is the least upper bound of the domain of $x(t, \bar{t}, b)$, then by Extension Theorem 1.1 in Chapter 1,

$$\lim_{t \uparrow t_2} |x(t, \bar{t}, b)| = a \tag{2}$$

If $t \in [\bar{t}, t_2]$

$$V[t, x(t, \bar{t}, b)] - V[\bar{t}, b] = \int_{\bar{t}}^{t} \dot{V}[s, x(s, \bar{t}, b)] ds \leq 0$$

and hence

$$V[t, x(t, \bar{t}, b)] \leq V[\bar{t}, b]$$

Let $\varepsilon > 0$ be such that $\varepsilon < A$ and let

$$I = \{x / \varepsilon \leq |x| \leq A \leq a\}$$

Then I is compact, and $0 \notin I$. Hence

$$\lim_{x \in I} W(x) = \mu > 0$$

Since $V(t, 0, \ldots, 0) = 0$ for $t > t_1$, there is a number $\lambda > 0$ such that:

(i) $\lambda < \mu$

(ii) $|x| < \lambda$ implies $V(\bar{t}, x) < \mu$

Let b be such that $|b| < a$ and $|b| < \lambda$. Then if $t \in [\bar{t}, t_2)$

$$\mu > V(\bar{t}, b) \geq V(t, x(t, \bar{t}, b)) \geq W[x(t, \bar{t}, b)]$$

Then from the definition of μ, it follows that if $t \in [\bar{t}, t_2)$, then

$$|x(t, \bar{t}, b)| < \varepsilon < a \tag{3}$$

If $t_2 < \infty$, then (2) holds and (3) contradicts (2). Hence, $t_2 = \infty$. Thus if $|b| < \min(a, \lambda)$, then (3) holds for all $t > \bar{t}$. This completes the proof of the Stability Theorem. □

Asymptotic Stability Theorem. *If there exists a function $V(t, x)$ which is positive definite on S, if $\dot{V}(t, x)$ has an infinitesimal upper bound and if $\dot{V}(t, x)$ is negative definite on S, then $x(t) \equiv 0$ is an asymptotically stable solution of (1).*

Remark. The geometric meaning of this theorem is very similar to that of the Stability Theorem. The additional hypotheses that $\dot{V}(t, x)$ has an infinitesimal upper bound and that $\dot{V}(t, x)$ is negative definite rather than just negative semidefinite force a solution not only to stay inside an ellipsoid but to enter smaller ellipsoids.

Proof. From the Stability Theorem, solution $x(t) \equiv 0$ is stable. Hence, there is a positive constant L such that if $|b| < L$, then $x(t, \bar{t}, b)$ is defined for all $t > \bar{t}$ and $|x(t, \bar{t}, b)| < a$ for all $t > \bar{t}$. We complete the proof of the theorem by showing that if $|b| < L$,

$$\lim_{t \to \infty} |x(t, \bar{t}, b)| = 0$$

First suppose there exists a solution $\bar{x}(t, \bar{t}, b)$ such that $|b| < L$ and such that there exists a constant $m > 0$ and a number $t_0 > \bar{t}$ so that for all $t \geq t_0$

$$V[t, \bar{x}(t, \bar{t}, b)] \geq m > 0$$

Since V has an infinitesimal upper bound there is a number $h \in (0, a)$ such that if $|x| < h$, then $V(t, x) < m$ and hence for all $t > t_0$,

$$|\bar{x}(t, \bar{t}, b)| \geq h$$

By hypothesis, there exist positive definite functions $W(x)$, $W_1(x)$ such that if $t > t_1$ and $|x| < a$, then

$$V(t, x) \geq W(x)$$

and

$$-\dot{V}(t, x) \geq W_1(x)$$

Let

$$\mu_1 = \min_{|x| \in [h, A]} W_1(x)$$

Then for all $t \geq t_0$

$$W_1[\bar{x}(t, \bar{t}, b)] \geq \mu_1 > 0$$

and hence

$$-\dot{V}[t, \bar{x}(t, \bar{t}, b)] \geq W_1[\bar{x}(t, \bar{t}, b)] \geq \mu_1$$

and

$$\int_{t_0}^{t} \dot{V}[s, x(s, \bar{t}, b)] ds \leq \int_{t_0}^{t} (-\mu_1) ds$$

so that for all $t \geq t_0$,

$$V[t, \bar{x}(t, \bar{t}, b)] \leq V[t_0, \bar{x}(t_0, \bar{t}, b)] - \mu_1(t - t_0)$$

Since V is bounded, it follows that if t is sufficiently large,

$$V[t, \bar{x}(t, \bar{t}, b)] < 0$$

This contradicts the hypothesis that $V(t, x)$ is positive definite.

Hence, if $x(t, \bar{t}, b)$ is a solution such that $|b| < L$ and if $\delta > 0$, then there exists $t' > \bar{t}$ such that

$$V[t', x(t', \bar{t}, b)] < \delta \tag{4}$$

Also, if $\tilde{t}, \tilde{t}_0 \in [t_1, \infty)$ and $\tilde{t} < \tilde{t}_0$, then since $\dot{V}(t, x)$ is negative definite,

$$V[\tilde{t}_0, x(\tilde{t}_0, \tilde{t}, b)] - V[\tilde{t}, x(\tilde{t}, \tilde{t}, b)] = \int_{\tilde{t}}^{\tilde{t}_0} \dot{V}[s, x(s, \tilde{t}, b)] ds \leq 0$$

That is, $V[t, x(t, \tilde{t}, b)]$ is monotonic nonincreasing. This fact, combined with (4), shows that

$$\lim_{t \to \infty} V[t, x(t, \tilde{t}, b)] = 0$$

and

$$\lim_{t \to \infty} W[x(t, \tilde{t}, b)] = 0$$

Now, given $\delta \in (0, A)$, let

$$\mu = \min_{|x| \in [\delta, A]} W(x)$$

There exists T such that if $t > T$, then

$$W[x(t, \tilde{t}, b)] < \frac{\mu}{2}$$

Hence, $t > T$ implies

$$|x(t, \tilde{t}, b)| < \delta$$

This completes the proof of the Asymptotic Stability Theorem.   □

Instability Theorem. *Suppose there exists a function $V(t, x)$ with domain S which satisfies the following hypotheses:*

(1) *$V(t, x)$ has an infinitesimal upper bound;*

(2) *$\dot{V}(t, x)$ is positive definite on S;*

(3) *there exists $T > t_1$ such that if $\tilde{t} \geq T$ and h is a positive constant, then there exists $c \in R^n$ such that $|c| < h$ and such that $V(\tilde{t}, c) > 0$.*

Then $x(t) \equiv 0$ is not stable.

Proof. If $x(t)$ is a solution which is not identically zero, it follows by uniqueness of solution that for all $t > t_1$ for which $x(t)$ is defined

$$|x(t)| \neq 0$$

Hence if $x(t)$ is defined for $t = t_0$, then for all $t \geq t_0$ for which $x(t)$ is defined,

$$V[t, x(t)] - V[t_0, x(t_0)] = \int_{t_0}^{t} \dot{V} ds > 0 \tag{5}$$

Let $t_0 > T$ and let $\varepsilon > 0$. By hypothesis, there exists c such that

$$|c| < \min(a, \varepsilon)$$

and such that

$$V(t_0, c) > 0 \tag{6}$$

Let $x(t) = x(t, t_0, c)$. We will complete the proof of the theorem by showing that there is a finite value $\tau > t_0$ such that the domain of $x(t, t_0, c)$ does not contain τ. Since V has infinitesimal upper bound, there exists $\lambda \in (0, a)$ such that if $|x| < \lambda$ and $t > t_1$, then

$$|V(t, x)| < V(t_0, c) \tag{7}$$

If solution $x(t) = x(t, t_0, c)$ is defined for $t > t_0$, then by (5) and (6),

$$V[t, x(t)] > V[t_0, x(t_0)] > 0$$

It follows from (7) that for $t > t_1$,

$$|x(t)| \geq \lambda$$

By hypothesis, there is a positive definite function $W(x)$ such that for all $(t, x) \in S$, $\dot{V}(t, x) \geq W(x)$. Let

$$\mu = \min_{|x| \in [\lambda, a]} W(x)$$

Then

$$\dot{V}[t, x(t)] \geq W[x(t)] \geq \mu \tag{8}$$

Since $V(t, x)$ is bounded, i.e., there exists $L > 0$ such that for all $(t, x) \in S$, $V(t, x) < L$, then by (5) and (8)

$$L > V[t, x(t)] \geq V[t_0, x(t_0)] + \mu(t - t_0)$$

Thus the set of values $t > t_0$ for which $x(t)$ is defined has a finite upper bound. This completes the proof of the Instability Theorem. $\square$

Lyapunov functions can also be used to investigate other kinds of stability. We describe its use for two such kinds of stability.

Definition. Given the differential equation

$$x' = f(t, x) \tag{9}$$

and the n-vector function $U(t, x)$ where f and U have continuous first derivatives with respect to t and the components of x for all $(t, x) \in S$ where

$$S = \{(t, x) \mid t > -\delta, \ |x| < a\}$$

and $\delta > 0$; suppose also that for all $t > -\delta$,

$$f(t, 0) = 0$$

Then the solution $x(t) \equiv 0$ of equation (9) is *stable under persistent disturbances* if $\varepsilon > 0$ implies there exist positive numbers $d_1(\varepsilon)$ and $d_2(\varepsilon)$ such that if for $t > -\delta$ and $|x| < \varepsilon$,

$$|U(t, x)| < d_1(\varepsilon)$$

and if $x(t)$ is a solution of

$$x' = f(t, x) + U(t, x) \tag{10}$$

such that $|x(0)| < d_2(\varepsilon)$, then $x(t)$ is defined for all $t > 0$ and $|x(t)| < \varepsilon$ for all $t > 0$.

Note the similarity of this concept to the definition of structural stability.

Stability Under Persistent Disturbances Theorem. *Suppose there is a function $V(t,x)$ with domain S with the following properties:*

(1) *there exist positive definite functions $W(x)$, $W_1(x)$ such that if $(t,x) \in S$,*
$$W(x) \leq V(t,x) \leq W_1(x);$$

(2) *there exists a positive definite function $W_2(x)$ such that if $(t,x) \in S$,*
$$\dot{V} = \nabla V \cdot f + V_t \leq -W_2(x);$$

(3) *there exists a positive constant M such that if $(t,x) \in S$, then*
$$\left| \frac{\partial V}{\partial x_i}(t,x) \right| \leq M \qquad (i = 1, \ldots, n)$$

Then the solution $x(t) \equiv 0$ of (9) is stable under persistent disturbances.

Proof. Let $\varepsilon < a$ where a is the number which appears in the definition of the set S. Let
$$m = \min_{|x|=\varepsilon} W(x)$$

Then there exists $r \in (0,1)$ such that
$$\min_{r\varepsilon \leq |x| \leq \varepsilon} W(x) \geq \frac{3}{4} m$$

Since $W_1(0) = 0$ and W_1 is continuous, there exists a number $d \in (0, \varepsilon)$ such that if $|x| \leq d$, then
$$W_1(x) \leq \frac{m}{2}$$

Let μ be a number such that
$$0 < \mu < \min \left[W_2(x)/d \leq |x| \leq \varepsilon \right]$$

Let $k \in (0,1)$ and let

$$d_1(\varepsilon) = \frac{k\mu}{nM}$$

Now consider a solution $x(t)$ of (10) and suppose $|x(0)| < d$. If $|x(t)| < d$ for all $t > 0$ for which $x(t)$ is defined, then since $0 < d < a$, it follows (by Extension Theorem 1.1) that $x(t)$ is defined for all $t > 0$ and the proof is complete. Otherwise let t_1 be such that

$$|x(t_1)| = d$$

Then if $\bar{t} > t_1$ and if for $s \in [t_1, \bar{t}]$, $x(s)$ is defined and

$$d \le |x(s)| \le \varepsilon$$

we obtain: at any point $(s, x(s))$

$$
\begin{aligned}
V_t + (\nabla V) \cdot (f + U) &= V_t + \nabla V \cdot f + \nabla V \cdot U \\
&= \dot{V} + \nabla V \cdot U \\
&\le -W_2 + \nabla V \cdot U \\
&< -\mu + nM \left(\frac{k\mu}{nM} \right) = -\mu(1 - k) < 0
\end{aligned}
$$

Therefore

$$V\left[\bar{t}, x(\bar{t})|\right] \le V\left[t_1, x(t_1)\right] \le W_1\left[x(t_1)\right] \le \frac{m}{2} \tag{11}$$

Also

$$|x(\bar{t})| < r\varepsilon \tag{12}$$

because if

$$|x(\bar{t})| \ge r\varepsilon$$

then

$$V\left[\bar{t}, x(\bar{t})\right] \ge W\left[x(\bar{t})\right] \ge \frac{3}{4}m$$

which contradicts (11).

Now suppose it is not true that

$$|x(t)| < \varepsilon$$

for all $t > t_1$. Then there exists $\tilde{t} > t_1$ such that

$$|x(\tilde{t})| = \varepsilon$$

Let

$$s_1 = \min[t/|x(t)| = \varepsilon]$$
$$s_2 = \max[t/t < s_1 \text{ and } |x(t)| = d]$$

Then $s_2 < s_1$ and if $s \in [s_2, s_1]$, then

$$d \leq |x(s)| \leq \varepsilon$$

But then by the same argument used to obtain (12),

$$|x(s_1)| < r\varepsilon$$

This contradicts the definition of s_1. Hence for all $t > 0$,

$$|x(t)| < \varepsilon$$

With $d_2(\varepsilon) = d$, the theorem is proved. □

Now we consider equation (1) again but now let the set S be defined by:

$$S = \{(t, x)/t_1 < t, \ |x| < \infty\}$$

As before, we assume that $x(t) \equiv 0$ is a solution of (1). A function $V(t, x)$ is defined to be positive [negative] semidefinite and positive [negative] definite on the set S exactly as earlier.

Definition. Suppose that for each (t_0, x^0) such that $t_0 > t_1$ and $x^0 \in R^n$, the solution $x(t, t_0, x^0)$ of (1) is defined for all $t > t_0$, and that, given $\varepsilon > 0$, there exists $\delta > 0$ such that if

$$|x(\bar{t}, t_0, x^0)| < \delta$$

where $\bar{t} > t_1$, then for all $t > \bar{t}$,

$$|x(t, t_0, x_0)| < \varepsilon$$

and that

$$\lim_{t \to \infty} x(t, t_0, x^0) = 0$$

Then the solution $x(t) \equiv 0$ of (1) is *globally asymptotically stable.*

Global Asymptotic Stability Theorem. *Suppose that there exists a function $V(t, x)$ with domain S such that $V(t, x)$ has the following properties:*

(1) $V(t, x)$ has an infinitesimal upper bound on each set $S_1 = \{(t, x)/t_1 < t, \ |x| < a\}$ where a is a positive constant;

(2) $\dot{V}(t, x)$ is negative definite on S;

(3) $V(t, x)$ is positive definite on S, i.e., there is a function $W(x)$ such that $W(0) = 0$, $W(x) > 0$ if $|x| \neq 0$ and for all $(t, x) \in S$,

$$V(t, x) \geq W(x);$$

(4) $\lim_{|x| \to \infty} W(x) = \infty.$

Then $x(t) \equiv 0$ is a globally asymptotically stable solution of (1).

Proof. Let $x(t) = x(t, \bar{t}, b)$ be a solution of (1) such that $\bar{t} > t_1$ and $b \neq 0$. Since $\dot{V}(t, x)$ is negative semidefinite, then as shown in the proof of the Stability Theorem, the function $V[t, x(t)]$ is monotonic nonincreasing. Hence for $t \geq \bar{t}$,

$$W[x(t)] \leq V[t, x(t)] \leq V[\bar{t}, b]$$

Hence from condition (4), it follows that the set

$$\{x(t)/t \geq \bar{t}\}$$

is bounded. Hence $x(t)$ is defined for all $t \geq \bar{t}$ (by Exercise 15, Chapter 1). As in the proof of the Asymptotic Stability Theorem, it follows that

$$\lim_{t \to \infty} |x(t)| = 0 \qquad \square$$

Application of the Second Method

As an application of the Lyapunov second method, we give another proof of the Asymptotic Stability Theorem for Nonlinear Systems in Chapter 4. For this proof, we need the following lemma.

Lemma 5.1 *If A is a real $n \times n$ matrix and γ is a non-zero real number, there is a real nonsingular matrix P such that*

$$PAP^{-1} = \begin{bmatrix} D_1 & & & \\ & D_2 & & \\ & & \ddots & \\ & & & D_m \end{bmatrix}$$

where each D_j is a real square matrix associated with an eigenvalue λ_j of A. If λ_j is real,

$$D_j = \begin{bmatrix} \lambda_j & & & \\ \gamma & \lambda_j & & \\ & \gamma & & \\ & & \ddots & \\ & & \gamma & \lambda_j \end{bmatrix} \tag{13}$$

and if $\lambda = \alpha_j + i\beta_j$,

$$D_j = \begin{bmatrix} \alpha_j & \beta_j & & & & & \\ -\beta_j & \alpha_j & & & & & \\ \gamma & 0 & \alpha_j & \beta_j & & & \\ 0 & \gamma & -\beta_j & \alpha_j & & & \\ & & & & \ddots & & \\ & & & & \gamma & 0 & \alpha_j & \beta_j \\ & & & & 0 & \gamma & -\beta_j & \alpha_j \end{bmatrix} \tag{14}$$

Proof. From the real canonical form, it follows that it is sufficient to prove the lemma for a $q \times q$ matrix of the form

$$
C = \begin{bmatrix} \lambda & 1 & & \\ & \ddots & \ddots & \\ & & & 1 \\ & & & \lambda \end{bmatrix}
$$

where λ is real or a $2q \times 2q$ matrix of the form

$$
D = \begin{bmatrix}
\alpha & \beta & 1 & 0 & & & \\
-\beta & \alpha & 0 & 1 & & & \\
& & \alpha & \beta & & & \\
& & -\beta & \alpha & & 1 & 0 \\
& & & & & 0 & 1 \\
& & & & \ddots & \alpha & \beta \\
& & & & & -\beta & \alpha
\end{bmatrix}
$$

where α, β are real.

First, let

$$
R = \begin{bmatrix}
0 & \cdots & 0 & 1 \\
0 & \cdots & 1 & 0 \\
& & & \vdots \\
1 & \cdots & & 0
\end{bmatrix}
$$

Then $R^{-1} = R$ and

$$
RCR^{-1} = \begin{bmatrix}
\lambda & 0 & & \cdots & 0 \\
1 & \lambda & & \cdots & 0 \\
0 & 1 & \lambda & \cdots & 0 \\
& & \ddots & & \ddots \\
& & & 1 & \lambda
\end{bmatrix}
$$

Next let

$$S = \begin{bmatrix} 1 & 0 & \cdots & & 0 \\ 0 & \gamma & \cdots & & 0 \\ \cdot & & & & \cdot \\ 0 & \cdot & & \cdots & \gamma^{q-1} \end{bmatrix}$$

Then

$$S^{-1} = \begin{bmatrix} 1 & 0 & & \\ 0 & \frac{1}{\gamma} & & \\ & & \ddots & \\ & & & \frac{1}{\gamma^{q-1}} \end{bmatrix}$$

and

$$SRCR^{-1}S^{-1}$$

$$= \begin{bmatrix} 1 & & & \\ & \gamma & & \\ & & \ddots & \\ & & & \gamma^{q-1} \end{bmatrix} \begin{bmatrix} \lambda & 0 & \cdots & 0 \\ 1 & \lambda & & \\ & \ddots & \ddots & \\ & & 1 & \lambda \end{bmatrix} \begin{bmatrix} 1 & & & \\ & \frac{1}{\gamma} & & \\ & & \ddots & \\ & & & \frac{1}{\gamma^{q-1}} \end{bmatrix}$$

$$= \begin{bmatrix} \lambda & 0 & \cdots & & 0 \\ \gamma & \gamma\lambda & 0 \cdots & & 0 \\ & \gamma^2 & \gamma^2\lambda & \cdots & 0 \\ & & & \gamma^{q-1} & \gamma^{q-1}\lambda \end{bmatrix} \begin{bmatrix} 1 & & & \\ & \frac{1}{\gamma} & & \\ & & \ddots & \\ & & & \frac{1}{\gamma^{q-1}} \end{bmatrix}$$

$$= \begin{bmatrix} \lambda & & & \\ \gamma & \lambda & & \\ & \gamma & \lambda & \\ & & \gamma & \lambda \end{bmatrix}$$

To treat the matrix D: if R is a $2q \times 2q$ matrix, then

$$RDR^{-1} = \begin{bmatrix} \alpha & -\beta & & & & & & & \\ \beta & \alpha & & & & & & & \\ 1 & 0 & \alpha & -\beta & & & & & \\ 0 & 1 & \beta & \alpha & & & & & \\ & & 1 & 0 & \alpha & -\beta & & & \\ & & 0 & 1 & \beta & \alpha & & & \\ & & & & & & \cdot & \cdot & \cdot & \cdot \\ & & & & & & 1 & 0 & \alpha & -\beta \\ & & & & & & 0 & 1 & \beta & \alpha \end{bmatrix}$$

Let S be the $2q \times 2q$ matrix:

$$S = \begin{bmatrix} I_2 & & & \\ & \gamma I_2 & & \\ & & \ddots & \\ & & & \gamma^{q-1} I_2 \end{bmatrix}$$

where I_2 is the 2×2 identity matrix. $\qquad \square$

We let

$$V(t, x) = \sum_{i=1}^{n} x_i^2$$

Then $V(t, x)$ is certainly positive definite on (t, x)-space and has an infinitesimal upper bound. Hence, by the Asymptotic Stability Theorem in this chapter, it is sufficient to show that $\dot{V}(t, x)$ is negative definite for all x such that $|x| < a$, where a is some positive number. We prove this for the case where matrix A is a D_j as described in (13) and for the case where D_j is described as in (14). Suppose first that D_j has the form (13) and suppose $\lambda_j < -\sigma < 0$. Then

$$\dot{V} = 2x_1 x_1' + \cdots + 2x_n x_n'$$
$$= 2x_1 \lambda_j x_1 + 2x_2(\gamma x_1 + \lambda_j x_2) + \cdots + 2x_n(\gamma x_{n-1} + \lambda_j x_n)$$
$$\quad + 2x_1 h_1 + \cdots + 2x_n h_n$$

where $h(t, x) = (h_1(t, x), \ldots, h_n(t, x))$.

We may choose $\gamma \in (0, \sigma)$. Then

$$\dot{V} \leq -\sigma(x_1^2 + \cdots + x_n^2) + 2x_1 h_1 + \cdots + 2x_n h_x$$

Since $|h(t,x)| = o(|x|)$, then if $\sum\limits_{j=1}^{n} x_j^2$ is sufficiently small,

$$\dot{V} \le -\frac{\sigma}{2}(x_1^2 + \cdots + x_n^2).$$

If D_j has the form (14), let us, for convenience, delete the subscripts and write $\lambda_j = \alpha + i\beta$. Then

$$\begin{aligned}
\dot{V} &= 2x_1(\alpha x_1 - \beta x_2) + 2x_2(\beta x_1 + \alpha x_2) + 2x_3(\gamma x_1 + \alpha x_3 - \beta x_4) \\
&\quad + 2x_4(\gamma x_2 + \beta x_3 + \alpha x_4) + \cdots + 2x_{n-1}(\gamma x_{n-3} + \alpha x_{n-1} - \beta x_n) \\
&\quad + 2x_n(\gamma x_{n-2} + \beta x_{n-1} + \alpha x_n) + 2x_1 h_1 + \cdots + 2x_n h_n \\
&= 2\alpha(x_1^2 + x_2^2 + \cdots + x_n^2) + 2\gamma x_1 x_3 + 2\gamma x_2 x_4 + \cdots + 2\gamma x_{n-3} x_{n-1} \\
&\quad + 2\gamma x_{n-2} x_n + 2x_1 h_1 + \cdots + 2x_n h_n
\end{aligned}$$

If $\alpha < -\sigma < 0$, then

$$\begin{aligned}
\dot{V} \le {} &-2\sigma(x_1^2 + \cdots + x_n^2) + \gamma(x_1^2 + x_3^2) + \gamma(x_2^2 + x_4^2) \\
&+ \cdots + \gamma(x_{n-2}^2 + x_n^2) + 2x_1 h_1 + \cdots + 2x_n h_n
\end{aligned}$$

If we choose $\gamma \in (0,\sigma)$, then

$$\dot{V} \le -\sigma(x_1^2 + \cdots + x_n^2) + 2x_1 h_1 + \cdots + 2x_n h_n$$

Since $|h(t,x)| = o(|x|)$, then if $x_1^2 + \cdots + x_n^2$ is sufficiently small,

$$\dot{V} \le -\frac{\sigma}{2}(x_1^2 + \cdots + x_n^2). \qquad \square$$

Exercises for Chapter 5

1. Show that the function

$$V(t,x) = x_1^2 + (\sin^2 t)x_2^2$$

is positive semidefinite but not positive definite.

2. Show that

$$V(t,x) = x_1^2 + (1 + \sin^2 t)x_2^2$$

is positive definite and has an infinitesimal upper bound.

3. Show that

$$V(t, x) = \sin(tx)$$

is bounded but does not have an infinitesimal upper bound.

4. Show that $(0,0,0)$ is a globally asymptotically stable solution of

$$x' = -x - xy^2 - x^3$$
$$y' = -7y + 3x^2y - 2yz^2 - y^3$$
$$z' = -5z + y^2z - z^3$$

Hint: Use the Global Asymptotic Stability Theorem with

$$V(t, x) = 3x^2 + y^2 + 2x^2.$$

5. Given the n-dimensional system

$$x' = Ax + f(x) \qquad\qquad (*)$$

where the matrix A is such that each eigenvalue of A has negative real part, show that there exists $b > 0$ such that if

$$|f(x)| < b|x|$$

for all x with $x \neq 0$, then 0 is a globally asymptotically stable solution of $(*)$.

6. Prove the Instability Theorem for Nonlinear Systems (Chapter 4).

Chapter 6

Periodic Solutions

Introduction

In this chapter we have two objectives: first, to present some theorems concerning periodic solutions of nonlinear systems of ordinary differential equations, and, secondly, to determine which of these theorems is useful for applications. Thus, part of this chapter is a continuation of the general study of ordinary differential equations and part is directed toward applications.

The problem of studying periodic solutions is very old. It first became prominent in the study of celestial mechanics and there is a well-known discussion of its importance by Poincaré [1892-1899, volume I, pp. 81-82] which ends with the oft-quoted remark:

> "D'ailleurs, ce qui nous rend ces solutions périodiques si précieuses, c'est qu'elles sont, pour ainsi dire, la seule brèche par où nous puissons essayer de pénétrer dans une place jusqu'ici réputée inabordable."

In later years, when other physical phenomena have been studied mathematically by using nonlinear ordinary differential equations, periodic solutions have often played an important role: radio circuits, control theory, and, most recently, chemical and biological oscillations. Periodic solutions have been studied for many years and

217

in connection with many applications and, consequently, the literature on the subject is enormous. The results which we will describe
in this chapter and in Chapter 7 are concerned with the theoretical
aspects of just two kinds of results. In this chapter, we study the
Poincaré-Bendixson theorem and an important n-dimensional analog
of it due to Sell. In Chapter 7 we study the branching or bifurcation
of periodic solutions, i.e. the appearance of periodic solutions when
a small parameter is varied. In both chapters we place considerable
emphasis on stability properties of the solutions for reasons which
will be discussed in detail in this chapter.

The two kinds of results we discuss are the basis for a great many
studies of periodic solutions. However, it should be emphasized that
there are many other very important studies of periodic and almost
periodic solutions with which we will not deal at all. Among these
are the remarkable and famous results of Kolmogoroff, Arnold and
Moser. See Moser [1973]. One reason for not dealing with this important subject is that systems of differential equations which describe
biological or chemical situations can be expected to have some kind
of dissipative properties. See Moser [1973, page 40].

Poincaré-Bendixson Theorem

We begin with an old and widely applied result, the Poincaré-Bendixson Theorem. Very roughly, this theorem says that if a 2-dimensional
autonomous system has a solution which stays in a bounded region
and does not approach an equilibrium point, then the solution is itself periodic or it spirals toward a solution which is periodic. From
an intuitive viewpoint this is a rather reasonable result. Since the
solution stays in a bounded region and does not approach an equilibrium point, then it has to "pile up" some place; so it piles up on a
periodic solution. However, a rigorous proof of the theorem is fairly
lengthy and requires the full force of the Jordan Curve Theorem. As
we will see, part of the difficulty in proving the theorem lies in the
fact that all the considerations take place in the Euclidean plane and
distinguishing between intuitive and rigorous arguments in the plane
is sometimes difficult.

We start with a precise statement of the Poincaré-Bendixson Theorem. For this, we need the notion of the ω-limit set of a solution S, denoted by $\Omega(S)$. We remind the reader that if the solution S is bounded, then $\Omega(S)$ is nonempty, bounded, connected, closed and invariant (Theorem 3.4). We also use the notation $O(S)$ to denote the orbit of solution S.

Poincaré-Bendixson Theorem. *Given the autonomous system*

$$x' = P(x, y)$$
$$y' = Q(x, y) \tag{1}$$

where P, Q are continuous and satisfy a local Lipschitz condition at each point of an open set in R^2, suppose that the solution $S = (x(t), y(t))$ of (1) is defined for all $t \geq t_0$, where t_0 is a fixed value, and is such that there exists a number M such that for all $t \geq t_0$

$$|x(t)| + |y(t)| < M$$

Suppose also that $\Omega(S)$ contains no equilibrium points of (1). Then one of the following two alternatives holds. Either:

(1) *$(x(t), y(t))$ is a periodic solution (in which case $O(S) = \Omega(S)$);*

or

(2) *$\Omega(S)$ is the orbit of a periodic solution and solution S approaches $\Omega(S)$ "spirally from the inside" or "spirally from the outside." (The sense in which the words in quotes are used will be described in the proof of the theorem.)*

Definition. The orbit $\Omega(S)$ in alternative (2) is called a *limit cycle*.

In order to prove the Poincaré-Bendixson Theorem, we need several preliminary results.

Jordan Curve Theorem. *Let C be a simple closed curve in R^2. Then*

$$R^2 - C = O_1 \cup O_2$$

where O_1, O_2 are disjoint nonempty connected open sets such that:

(1) *For $i = 1, 2$, the boundary of O_i is C.*

(2) *One of the open sets, say O_1, is bounded (it is called the* interior *of C) and the other, O_2, is unbounded (it is called the* exterior *of C).*

(3) *If $p \in O_2$, then*

$$i(C, p) = 0$$

For all $p \in O_1$, the index $i(C,p)$ has the same value, either $+1$ or -1. (The sign depends on the orientation assigned to C.)

A rigorous definition of index and a reference for a proof of the Jordan Curve Theorem are given in the Appendix.

Definition. A point $(x_0, y_0) \in D$ which is not an equilibrium point of (1) is a *regular point*.

Definition. A *finite closed segment of a straight line* is a set of points of one of the following forms:

$$L = \{(x,y)/y = mx + b \quad \text{and} \quad c \le x \le d\}$$

or

$$L = \{(e,y)/f \le y \le g\}$$

where m, b, c, d, e, f, g are constants.

Definition. Let $V = V(x,y)$ denote the vector field $(P(x,y),Q(x,y))$ with domain D. A *transversal* (or *segment without contact*) of V is a finite closed segment of a straight line, say L, such that

(1) $L \subset D$;

(2) If $(x,y) \in L$, then (x,y) is a regular point of V;

(3) If $(x_1, y_1) \in L$, then the slope of $V(x_1,y_1)$ is not equal to the slope of L, i.e. L is not tangent to an orbit of (1).

We list some properties of transversals which will be needed.

(1) If (x_0, y_0) is a regular point of $V(x, y)$ and if λ is a line which contains (x_0, y_0) and which is not parallel to $V(x_0, y_0)$, there exists a transversal $L \subset \lambda$ such that (x_0, y_0) is in the interior of L.

Proof. Since P, Q are continuous, there is a circular neighborhood $N(x_0, y_0)$ such that if $(x, y) \in \overline{N(x_0, y_0)}$ then the vector $(P(x, y), Q(x, y))$ is not parallel to λ. Let

$$L = \left\{ \overline{N(x_0, y_0)} \right\} \cap \lambda \qquad \square$$

(2) All orbits of (1) which intersect transversal L cross L in the same direction as t increases.

Proof. If orbits C_1 and C_2 cross L in different directions at points p_1 and p_2, then there exists $p_3 \in L$ such that p_3 is between p_1 and p_2 and such that the orbit through p_3 is tangent to L. (This follows from the continuity of P and Q and the Intermediate Value Theorem.)$\square$

(3) If
$$F = \{(x(t), y(t))/t \in [a, b]\}$$

is a finite arc of the orbit C of a solution of equation (1), and if L is a transversal, then F can cross L only a finite number of times.

Proof. Suppose this is not true. Then there exists a monotonic sequence $\{t_n\} \subset [a, b]$ such that

$$\lim_{n \to \infty} t_n = t_0 \in [a, b]$$

and $(x(t_n), y(t_n)) \in L$ for all n, and such that for all n,

$$(x(t_n), y(t_n)) \neq (x(t_0), y(t_0))$$

Let

$$A_0 = (x(t_0), y(t_0))$$

and

$$A_n = (x(t_n), y(t_n))$$

Since

$$\lim_{n\to\infty} A_n = A_0$$

then the limiting direction of the secant $\overline{A_0 A_n}$ of orbit C is the direction of the tangent to C at $(x(t_0), y(t_0))$. But for all n, $\overline{A_0 A_n}$ is contained in L. Hence L is tangent to C at $(x(t_0), y(t_0))$. This contradicts the condition that L is a transversal. $\qquad\square$

(4) Let A be an interior point of transversal L. Then, given $\varepsilon > 0$, there exists $r > 0$ such that if Δ is a disc with center A and radius less than or equal to r, then if orbit C (described by solution $(x(t), y(t))$) is in disc Δ at $t = 0$ (i,e. $(x(0), y(0)) \in \Delta$) there exists t_1 such that $|t_1| < \varepsilon$ and $(x(t_1), y(t_1)) \in L$. (See Figure 1.)

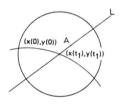

Figure 1

Proof. We assume that the coordinate axes have been rotated and translated so that A is the origin of the coordinate system and L is contained in the x-axis. From the Basic Existence Theorem 1.1, there is a unique solution $(x(t,0,0), y(t,0,0))$ of (1) such that

$$x(0,0,0) = 0, \quad y(0,0,0) = 0$$

and there exists a neighborhood N of $(0,0)$ such that if $(x_0, y_0) \in N$, then there is a solution $(x(t, x_0, y_0), y(t, x_0, y_0))$ such that

$$x(0, x_0, y_0 = x_0$$
$$y(0, x_0, y_0) = y_0$$

If

$$\frac{\partial}{\partial t} y(0,0,0) = 0$$

then the solution $(x(t,0,0), y(t,0,0))$ is tangent to the x-axis (Transversal L) at the origin A. This contradicts the fact that L is a transversal. Hence

$$\frac{\partial}{\partial t} y(0,0,0) \neq 0$$

and we can apply the Implicit Function Theorem to solve the equation

$$y(t, x_0, y_0) = 0$$

uniquely for t as a function of (x_0, y_0) in a neighborhood of $t = x_0 = y_0 = 0$. □

Now we obtain some lemmas which are needed to prove the Poincaré-Bendixson Theorem.

Lemma 6.1 *Suppose $\Omega(S)$ contains a regular point A, and let L be a transversal such that A is an interior point of L. Then there exists a monotonic sequence $\{t_m\}$ such that $t_m \to \infty$ and such that if*

$$A_m = (x(t_m), y(t_m))$$

then if L is contained in a sufficiently small neighborhood of A,

$$A \cup (\cup_m A_m) = [0(S)] \cap L$$

If $A_1 = A_2$, then $A = A_m$ for all m and $O(S)$ is a simple closed curve. If $A_1 \neq A_2$, then all the A_m's are distinct (i.e. if $i \neq j$, then $A_i \neq A_j$) and for all m, A_{m+1} is between A_m and A_{m+2} on L.

Proof. Note first that a transversal L exists by the first of the properties of transversals that were listed earlier. By property 4 of transversals with $\varepsilon = 1$ and from the fact that $A \in \Omega(S)$, it follows that there exists a monotonic squence $\{t_m\}$ such that $t_m \to \infty$ and for each m,

$$(x(t_m), y(t_m)) \in O(S) \cap L$$

By property 3 of transversals, the sequence $\{t_m\}$ can be chosen so that if

$$\bar{t} \in (t_m, t_{m+1})$$

then

$$(x(\bar{t}), y(\bar{t})) \notin L.$$

Let $A_m = (x(t_m), y(t_m))$. Then by (2), either $\{A_m\}$ is a finite set or A is a limit point of the set $\{A_m\}$. If $A_1 = A_2$, then by Lemma 3.3 of Chapter 3, $O(S)$ is a simple closed curve and $A_2 = A_3 = A_m = A$ for all m because if $s > t_2$,

$$(x(s), y(s)) \in \{(x(t), y(t))/t \in [t_1, t_2]\}$$

Now suppose $A_1 \neq A_2$. If $t \in (t_1, t_2)$, then $(x(t), y(t)) \notin L$. Hence the line segment $\overline{A_1 A_2}$ and the curve

$$\{(x(t), y(t))/t \in [t_1, t_2]\}$$

form a simple closed curve C.

The conventional argument for the proof of Lemma 6.1 then proceeds as follows. We consider two cases.

Case I. There exists $\varepsilon > 0$ such that if $t \in (t_2, t_2+\varepsilon)$ then $(x(t), y(t))$ is an element of the interior of C. (See Figure 2.) Then for all $t > t_2$, $(x(t), y(t))$ is an element of the interior of C. In order to show this, assume it is not true and let

$$t' = \text{g}\ell\text{b}\,\{t > t_2/(x(t), y(t)) \text{ is not in the interior of } C\}$$

Then $(x(t'), y(t')) \in C$. Since $O(S)$ cannot cross itself (by Theorem 3.2, Chapter 3), then $(x(t'), y(t'))$ is a point in the interior of the line segment $\overline{A_1 A_2}$. but then $(x(t), y(t))$ crosses transversal L at $(x(t'), y(t'))$ in the direction opposite to the direction of the crossing at $(x(t_2), y(t_2))$. This contradicts property 2 of transversals.

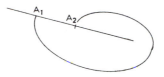

Figure 2

Next, $A_3 \neq A_2$ because otherwise $O(S)$ would intersect itself but would not be a closed curve. By the argument in the preceding paragraph, $A_3 \neq \overline{A_1 A_2}$. All points on L "to the left of" A_1 are in the exterior of C. Hence, A_3 is "to the right of" A_2. (The phrases enclosed in quotation marks can easily be replaced by rigorous language.) The remainder of the proof follows by induction since Case I also holds at A_3, i.e. there exists $\varepsilon > 0$ such that if $t \in (t_2, t_2 + \varepsilon)$ then $(x(t), y(t))$ is an element in the interior of the simple closed curve formed by $\overline{A_2 A_3}$ and the curve

$$\{(x(t), y(t))/t \in [t_2, t_3]\}$$

Case II. There exists $\varepsilon > 0$ such that if $t \in (t_2, t_2 + \varepsilon)$, then $(x(t), y(t))$ is an element of the exterior of C. (See Figure 3.) By the same kind of argument as for Case I, the set

$$\{(x(t), y(t))/t > t_2\}$$

is in the exterior of C. The remaining steps are also parallel to those in Case I except that at some stage, we may be reduced from Case II to Case I (as, for example in Figure 4). This completes the proof of Lemma 6.1. □

Figure 3

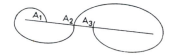

Figure 4

Notice that in this proof, we use only parts (1) and (2) in the statement of the Jordan Curve Theorem.

Now we indicate how to complete the proof of Lemma 6.1 starting from the beginning of Case I without splitting the proof into cases and without making any appeal to a geometric picture, i.e. Figures 2, 3, and 4. For this proof, we require part (3) of the statement of the Jordan Curve Theorem. Let C be the simple closed curve formed by the segment $\overline{A_1 A_2}$ and the curve

$$\{(x(t), y(t))/t \in [t_1, t_2]\}$$

If for some $t > t_2$, the solution crosses $\overline{A_1 A_2}$, then by property (2) of transversals, it must cross as indicated by the dashed arrow in Figure 5 i.e., in the same direction as the crossing at A_2. Let

$$\tilde{t} = \mathrm{lub}\{t > t_2\}/(x(s), y(s)) \notin \overline{A_1 A_2} \quad \text{for} \quad t_2 \leq s \leq t$$

and let

$$p = (x(t_0), y(t_0))$$

where

$$t_2 < t_0 < \tilde{t}$$

and let

$$\tilde{p} = (x(\tilde{t}), y(\tilde{t}))$$

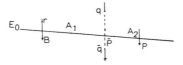

Figure 5

Then it is easy to show that there exist points q and $\bar{q}$, as shown in Figure 5, such that

$$i(C,q) \neq i(C,\bar{q}) \quad \text{(by the Jordan Curve theorem)} \tag{3}$$

and

$$i(C,p) = i(C,\bar{q}) \tag{4}$$

But

$$i(C,p) = i(C,q) \tag{5}$$

because the subset of the solution curve between p and q does not cross $\overline{A_1 A_2}$. Equations (4) and (5) contradict (3). Hence if $t > t_2$, the solution does not cross $\overline{A_1 A_2}$.

Let E_0 be the endpoint of L that is on the other side of A_1 from A_2. (It is straightforward to translate this statement into rigorous language.) It remains to show that for $t > t_2$, the solution curve does not cross the segment $\overline{E_0, A_1}$. Suppose the curve does cross $\overline{E_0 A_1}$. Let

$$t_3 = \min\left\{ t/t > t_2 \quad \text{and} \quad (x(t), y(t)) \in \overline{E_0 A_1} \right\}$$

and let

$$B = (x(t_3), y(t_3))$$

Since the crossing must take place in the direction indicated in Figure 5, then if the point r is as indicated in Figure 5,

$$i(C, p) = i(C, r) \tag{6}$$

(since the part of the orbit joining the points p and r does not intersect the curve C) and

$$i(C, r) = i(C, q) \tag{7}$$

if the segment rq is close enough to the segment $\overline{A_1 A_2}$. But by (3) and (4),

$$i(C, q) \neq i(C, p) \tag{8}$$

But (6), (7) and (8) yield a contradiction.   □

Lemma 6.2 *If L is a transversal of sufficiently short length, then $L \cap \Omega(S)$ contains at most one point.*

Proof. Since L is a transversal, L contains no equilibrium points. By Lemma 6.1, $L \cap \Omega(S)$ contains at most one point because if $L \cap \Omega(S)$ contains two points B_1 and B_2, then by Lemma 6.1 there exist sequences $\{A_m^{(1)}\}$ and $\{A_m^{(2)}\}$ such that $\lim A_m^{(1)} = B_1$ and $\lim A_m^{(2)} = B_2$, and

$$O(S) \cap L = \cup_m A_m^{(1)}$$

and

$$O(S) \cap L = \cup_m A_m^{(2)}$$

Since $B_1 \neq B_2$, then

$$\cup_m A_m^{(1)} \neq \cup_m A_m^{(2)}$$

We have obtained a contradiction to Lemma 6.1.   □

Lemma 6.3 *If $O(S)$ is a closed curve, then $O(S) = \Omega(S)$.*

Proof. First we show that $O(S) \supset \Omega(S)$. Since $O(S)$ is a closed curve, there exists numbers t_1, t_2 such that $t_1 < t_2$ and

$$O(S) = \{(x(t), y(t))/t \in [t_1, t_2]\}$$

Thus $O(S)$ is compact (because it is the continuous image of a compact set) and hence contains its limit points. But an ω-limit point of S is a limit point of $O(S)$.

Now we show that $O(S) \subset \Omega(S)$. Let T be the period of $(x(t), y(t))$, and let $(x(t_0), y(t_0)) \in O(S)$. Then

$$\lim_{m \to \infty} (x(t_0 + mT), y(t_0 + mT)) = \lim_{m \to \infty} (x(t_0), y(t_0)) = (x(t_0), y(t_0)) \quad \square$$

Lemma 6.4 *If* $\Omega(S) \cap O(S) \neq \phi$, *then* $O(S)$ *is a closed curve.*

Proof. If $A \in \Omega(S) \cap O(S)$ then A is a regular point because each point of $O(S)$ is a regular point. If $O(S)$ is not a closed curve, then by Lemma 6.1 there is a sequence of distinct points $\{A_n\}$ such that for all n, $A_n \in O(S) \cap L$ where L is a transversal which has A as an interior point. But $O(S) \subset \Omega(S)$ because $\Omega(S)$ is invariant (by Theorem 3.4). Hence

$$\{A_n\} \subset O(S) \cap L \subset \Omega(S) \cap L$$

This contradicts Lemma 6.2. $\quad \square$

Lemma 6.5 *If* $\Omega(S)$ *contains no equilibrium points and if* $\Omega(S) \supset O(S_1)$ *where* S_1 *is a periodic solution, then* $\Omega(S) \subset O(S_1)$.

Proof. Suppose the set

$$Q = [\Omega(S)] \cap [O(S_1)]^c$$

is nonempty. Since the set $O(S_1)$ is a closed set, the set Q is not closed because otherwise

$$\Omega(S) = O(S_1) \cup Q$$

is the union of two disjoint closed bounded sets and this contradicts the connectedness of $\Omega(S)$ (given by Theorem 3.4).

Next we show that there exists a limit point p of Q such that $p \in O(S_1)$ by the following argument: first, since $Q \neq \phi$ and is not closed, then Q is infinite. Also, Q is bounded because $\Omega(S)$ is bounded since solution S is bounded. Hence Q has a limit point, and since Q is not closed there is a limit point p such that $p \notin Q$. Hence

$$p \in Q^c = [\Omega(S)]^c \cup O(S_1)$$

Since p is a limit point of Q, then p is a limit point of $\Omega(S)$. But $\Omega(S)$ is closed (by Theorem 3.4). Hence $p \in \Omega(S)$. Since

$$p \in [\Omega(S)]^c \cup O(S_1)$$

it follows that $p \in O(S_1)$.

Let L be a transversal such that p is an interior point of L. If $N_\varepsilon(p)$ is a circular neighborhood of p of radius $\varepsilon > 0$, then the set

$$N_\varepsilon(p) \cap Q = [N_\varepsilon(p)] \cap \{\Omega(S) \cap [O(S_1)]^c\}$$

is nonempty (because p is a limit point of Q) and consists of regular points because p is regular. Let $q \in N_\varepsilon(p) \cap Q$. By property (4) of transversals, L is intersected at point $\bar{p}$ by an orbit $O(S_2)$ through q. $O(S_2)$ is contained in $\Omega(S)$ because $q \in \Omega(S)$ and $\Omega(S)$ is invariant (by Theorem 3.4). Since $q \in [O(S_1)]^c$, then

$$O(S_2) \cap O(S_1) = \phi$$

Hence since

$$p \in O(S_1) \cap L \subset \Omega(S) \cap L$$

and

$$\bar{p} \in O(S_2) \cap L \subset \Omega(S) \cap L$$

the points p and $\bar{p}$ are distinct points in $\Omega(S) \cap L$. this contradicts Lemma 6.2 and hence completes the proof of Lemma 6.5.   □

Proof of Poincaré-Bendixson Theorem.   Let $p \in \Omega(S)$. Since $\Omega(S)$ contains no equilibrium points, there is a solution $\bar{S}$ with orbit $O(\bar{S})$

such that $p \in O(\bar{S})$. Since $\Omega(S)$ is invariant, $O(\bar{S}) \subset \Omega(S)$. If $\bar{p} \in \Omega(\bar{S})$, then $\bar{p} \in O(\bar{S})$ or $\bar{p}$ is a limit point of $O(\bar{S})$. Hence, since $\Omega(S)$ is closed, then

$$\Omega(\bar{S}) \subset \Omega(S)$$

Also, since $\Omega(S)$ contains no equilibrium points, then $\bar{p}$ is regular. Now by property (1) of transversals, there is a transversal L such that $\bar{p}$ is an interior point of L, and by Lemma 6.2

$$L \cap \Omega(S) = \bar{p}$$

Since $O(\bar{S}) \subset \Omega(S)$, then $L \cap [O(\bar{S})]$ contains at most one point. Hence, by Lemma 6.1, $\bar{S}$ is periodic. Since $O(\bar{S}) \subset \Omega(S)$, then by Lemma 6.5,

$$\Omega(S) = O(\bar{S})$$

If S is periodic, then by Lemma 6.3

$$O(S) = \Omega(S) = O(\bar{S})$$

If S is not periodic, then by Lemma 6.4

$$\Omega(S) \cap O(S) = \phi$$

and since $O(\bar{S}) = \Omega(S)$, then

$$O(\bar{S}) \cap O(S) = \phi$$

$O(\bar{S})$ is a simple closed curve and $O(S)$ is connected. Hence, $O(S)$ is in the interior of $O(\bar{S})$ or in the exterior of $O(\bar{S})$. Let $q \in O(\bar{S})$. Since q is regular, there is a transversal L such that q is an interior point of L. Since S is not periodic, then by Lemma 6.1

$$[O(S)] \cap L = \{A_m\} \tag{9}$$

where $\{A_m\}$ is a sequence of distinct points linearly ordered on L by subscript and $\lim_{m \to \infty} A_m = q$. Also, since $O(\bar{S}) = \Omega(S)$, then if $\mathcal{U}$ is an open set such that

$$O(\bar{S}) \subset \mathcal{U}$$

then there is a number τ_0 such that if $t \geq \tau_0$, then

$$(x(t), y(t)) \in \mathcal{U} \qquad (10)$$

Solution S spirals toward $\Omega(S) = O(\bar{S})$ in the sense described by (9) and (10).

This completes the proof of the Poincaré-Bendixson Theorem. $\square$

Application of the Poincaré-Bendixson Theorem

Next we discuss some of the significance, limitations and applications of the Poincaré-Bendixson Theorem.

As pointed out earlier, the proof of the Poincaré-Bendixson Theorem depends heavily on use of the Jordan Curve Theorem, which is a theorem in the xy-plane. Also, the intuitive idea (described earlier), that a bounded solution would tend to "pile up" on a periodic solution, no longer has much validity if the solution has an n-dimensional space, where $n > 2$, in which to "move about." Consequently, it is natural to expect that there is no n-dimensional generalization of the Poincaré-Bendixson Theorem. Indeed, we can construct examples which show that there is no such n-dimensional generalization. (See Exercise 4.)

The Poincaré-Bendixson Theorem can be generalized to a statement about solutions of differential equations on 2-dimensional manifolds (see A. J. Schwartz [1963]). It is also important in the theory of dynamical systems. For example, Moser [1973, p. 109] points out that it is essentially a consequence of the Poincaré-Bendixson Theorem that the two-dimensional sphere, the projective plane and the Klein bottle do not admit ergodic flows.

We are chiefly interested in the Poincaré-Bendixson Theorem because it provides a means for establishing the existence of periodic solutions. First, we indicate methods for showing that the hypotheses of the Poincaré-Bendixson Theorem are satisfied. Then we discuss stability properties of limit cycles.

Often, instead of showing that a particular solution is bounded, it is shown that there exists a bounded open set such that no solution whose orbit contains a point in the bounded open set "escapes" the open set at a later value of t. For example, if the boundary of the open set $\mathcal{U}$ is a simple closed curve K which is such that each orbit which intersects K crosses K going inward, then each solution which passes through a point of $\mathcal{U}$ stays in $\mathcal{U}$ for all later t. (See Figure 6.) One way to obtain such a curve K is by using Lyapunov functions. For example, consider the system

$$x' = f(x, y) - x^{2p+1}$$
$$y' = g(x, y) - y^{2p+1}$$

where f and g have continuous first derivatives in both variables and are of order less than $2p + 1$ in $r = \sqrt{x^2 + y^2}$ as $r \to \infty$, i.e.

$$\lim_{r \to \infty} \frac{|f(x, y)|}{r^{2p+1}} = 0$$

and

$$\lim_{r \to \infty} \frac{|g(x, y)|}{r^{2p+1}} = 0$$

Let

$$V(x, y) = \frac{1}{2}(x^2 + y^2)$$

Then $V(x, y)$ is positive definite and there exists $R > 0$ such that if $x^2 + y^2 \geq R^2$, then

$$\dot{V}_t = \nabla V \cdot (f(x, y) - x^{2p+1}, \; g(x, y) - y^{2p+1})$$
$$= x f(x, y) + y g(x, y) - x^{2p+2} - y^{2p+2}$$

is negative. Hence, the circle $x^2 + y^2 = R^2$ is such a curve K.

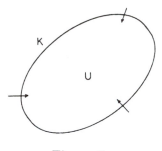

Figure 6

Generally, however, finding a curve K is not simple. Examples of such constructions for important special cases and references for other work in this direction are given by Lefschetz [1962, Chapter XI].

The problem of showing that $\Omega(S)$ contains no equilibrium points is often resolved by studying the equilibrium points themselves and showing that each is contained in the interior of a simple closed curve such that no solution enters that interior and hence is not an ω-limit point. For example, if (x_0, y_0) is an equilibrium point of

$$x' = P(x, y)$$
$$y' = Q(x, y)$$

and if the eigenvalues of the matrix

$$\begin{bmatrix} P_x(x_0, y_0) & P_y(x_0, y_0) \\ Q_x(x_0, y_0) & Q_y(x_0, y_0) \end{bmatrix}$$

both have positive real parts, then it is easy to show that (x_0, y_0) has that property. (See Exercise 1.) It must be emphasized that this procedure is not, in general, simple because if $P(x, y)$ and $Q(x, y)$ are complicated functions, determining the equilibrium points may require a nontrivial computation.

Now we study some of the stability properties of limit cycles, i.e. the periodic solutions which are the Ω-limit sets of certain bounded solutions. To see the variety of behavior that can occur, we first consider some examples.

Example 6.1 We consider:

$$x' = -y + \frac{x}{\sqrt{x^2 + y^2}}\left[1 - (x^2 + y^2)\right]$$
$$y' = x + \frac{y}{\sqrt{x^2 + y^2}}\left[1 - (x^2 + y^2)\right]$$

Using the polar coordinate $r = \sqrt{x^2 + y^2}$, we may write this example as:

$$x' = -y + \frac{x}{r}(1 - r^2) \tag{11}$$

$$y' = x + \frac{y}{r}(1 - r^2) \tag{12}$$

Since

$$rr' = xx' + yy'$$

then, multiplying (11) by x and (12) by y and adding, we obtain

$$rr' = r(1 - r^2)$$

or

$$r' = (1 - r^2) \tag{13}$$

Since

$$\theta' = \frac{xy' - yx'}{r^2}$$

then, multiplying (11) by $-y$ and (12) by x and adding and then dividing by r^2, we obtain:

$$\theta' = \frac{y^2 + x^2}{r^2} = 1$$

Now (13) can be written:

$$\frac{1}{2}\left\{\frac{r'}{1 - r} + \frac{r'}{1 + r}\right\} = 1$$

or

$$\frac{1}{2}\left\{\frac{dr}{1+r} + \frac{dr}{1-r}\right\} = dt$$

and integrating we obtain

$$\ell n\left|\frac{1+r}{1-r}\right| = 2t + C = 2t + \ell n\left|\frac{1+r_0}{1-r_0}\right|$$

where $r = r_0$ at $t = 0$. Then if $0 < r < 1$

$$\frac{1+r}{1-r} = \left[\frac{1+r_0}{1-r_0}\right]e^{2t}$$

or

$$r = \frac{Ke^{2t} - 1}{Ke^{2t} + 1} \tag{14}$$

where

$$K = \left[\frac{1+r_0}{1-r_0}\right]$$

If $r > 1$

$$r = \frac{Ke^{2t} + 1}{Ke^{2t} - 1} \tag{15}$$

where

$$K = \left[\frac{r_0 + 1}{r_0 - 1}\right]$$

Inspection of (14) and (15) show that $r \to 1$ as $t \to \infty$. Also, $r = 1$ is the orbit of a periodic solution. Hence, (14) shows that all solutions inside the circle $r = 1$ spiral toward the circle and all solutions outside the circle $r = 1$ spiral toward the circle. It is clear that $r = 1$ is asymptotically orbitally stable, and it is easy to show (Exercise 2) that $r = 1$ is phase asymptotically stable. Thus, the solution

$$x(t) = \cos[\theta(t)]$$
$$y(t) = \sin[\theta(t)]$$

where

$$\theta(t) = t$$

is clearly a limit cycle and, moreover, every solution approaches this limit cycle.

If all limit cycles had such strong stability properties, the Poincaré-Bendixson Theorem would be far more valuable in applied mathematics. Unfortunately, many limit cycles have no stability properties that have any physical significance, as we show now with examples.

Example 6.2

$$x' = \alpha \left(\sqrt{x^2 + y^2} \right) x + \beta y$$
$$y' = -\beta x + \alpha \left(\sqrt{x^2 + y^2} \right) y$$

where $\alpha(r)$ is a monotonic differentiable function for all $r > 0$ and

$$\alpha(r) = 0 \qquad r \le 1$$
$$\alpha(r) < 0 \qquad r > 1$$

(For an example of such a function, see Exercise 3.) If

$$V(x, y) = x^2 + y^2$$

then

$$\dot{V} = 2\alpha x^2 + 2\beta xy - 2\beta xy + 2\alpha y^2$$
$$= 2\alpha(x^2 + y^2)$$

where $\alpha = \alpha(r)$. Hence, every orbit which intersects a circle with center 0 and radius > 1 crosses the circle into its interior as t increases. On the other hand, as shown near the end of Chapter 3 (see Case V), every orbit which passes through a point in the interior of the circle with center 0 and radius 1 is a circle with center 0. Hence, the orbits are as sketched in Figure 7. The solution

$$x(t) = \sin \beta t$$
$$y(t) = \cos \beta t \tag{16}$$

is clearly a limit cycle. It is indeed the Ω-limit set of each solution which passes through a point outside the circle with center 0 and radius 1. This is intuitively quite clear, but the proof is slightly complicated by the fact that $\dot{V} = 0$ if $r = 1$. To give a precise proof, let $(\bar{x}(t), \bar{y}(t))$ be a solution which passes through a point outside the circle with center 0 and radius 1. If

$$V(x, y) = x^2 + y^2$$

then

$$\dot{V} = 2\alpha \left(\sqrt{\bar{x}^2 + \bar{y}^2} \right) ([\bar{x}(t)]^2 + [\bar{y}(t)]^2)$$

and hence $V[\bar{x}(t), \bar{y}(t)]$ is monotonic decreasing. Hence, if there exists a strictly increasing sequence $\{t_n\}$ such that $t_n \to \infty$ and such that

$$\lim_{n \to \infty} \{[\bar{x}(t_n)]^2 + [\bar{y}(t_n)]^2\} = B > 1$$

then for all t,

$$V[\bar{x}(t), \bar{y}(t)] = \{[\bar{x}(t)]^2 + [\bar{y}(t)]^2\} \geq B > 1 \qquad (17)$$

and there exists $M > 0$ such that for all t

$$\dot{V} \leq \alpha(B) \{[x(t)]^2 + [y(t)]^2\} < -M < 0$$

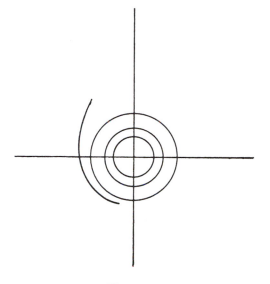

Figure 7

Let $\varepsilon > 0$ and suppose $\bar{t}$ is such that

$$\{[\bar{x}(\bar{t})]^2 + [\bar{y}(\bar{t})]^2\} < B + M\varepsilon$$

Let $\tilde{t}$ be such that $\tilde{t} - \bar{t} > 2\varepsilon$. Then

$$V[\bar{x}(\tilde{t}), \bar{y}(\tilde{t})] - V[\bar{x}(\bar{t}), \bar{y}(\bar{t})] = [\dot{V}(\tau)][\tilde{t} - \bar{t}] < -M(2\varepsilon)$$

where $\tau \in (\bar{t}, \tilde{t})$. Hence

$$V[\bar{x}(\tilde{t}), \bar{y}(\tilde{t})] < B + M\varepsilon - M(2\varepsilon) = B - M\varepsilon$$

This contradicts (17). Hence, if $\{t_n\}$ is a sequence such that $t_n \to \infty$, then

$$\lim_{n \to \infty} \{[\bar{x}(t_n)]^2 + [\bar{y}(t_n)]^2\} = 1$$

Thus the solution (16) is stable, but is certainly not asymptotically orbitally stable or phase asymptotically stable because of the behavior of the orbits in the interior of the circle with center 0 and radius 1.

If we were to investigate a physical system which was described by these equations, then if there were small disturbances of the system, the solutions of the equations could not be used to make definite predictions about the behavior of the physical system. It would be possible to predict that, after sufficient time had elapsed, both $x(t)$ and $y(t)$ would be less than or equal to one (the orbit would be in the unit circle), but no other predictions about values of $x(t)$ and $y(t)$ could be made. If a small disturbance shifted the physical system from one circular orbit to another, there would be no tendency for the physical system to return to the original orbit. Hence, if small disturbances occurred fairly frequently, the only prediction that could be made would be that $(x(t), y(t))$ would tend to remain in the unit circle. Certainly no prediction of periodicity could be made.

Example 6.3

$$x' = \alpha \left(\sqrt{x^2 + y^2} \right) x + \beta y$$
$$y' = -\beta x + \alpha \left(\sqrt{x^2 + y^2} \right) y$$

where $\alpha(r)$ is a differentiable function for $r > 0$ and

$$
\begin{aligned}
\alpha(r) &< 0 & r &> 1 \\
\alpha(r) &= 0 & r &= 1 \\
\alpha(r) &< 0 & r &< 1
\end{aligned}
$$

By the same kind of study as used in Example 2, it follows that the circle $r = 1$ is a limit cycle and is the Ω-limit set of every orbit which passes through a point which is outside the circle with center 0 and radius 1. On the other hand, by application of a familiar argument, it follows that every orbit which passes through a point in the interior of the circle with center 0 and radius 1 approaches the origin as $t \to \infty$. The limit cycle $r = 1$ is said in this case to be *semi-stable*.

Stable Periodic Solutions in the n-Dimensional Case

With these examples in mind, we consider the following question. Given a system of differential equations which describe a physical system, what kind of periodic solutions of the system of differential equations are reflected in oscillatory behavior of the physical system? That is, what properties should a periodic solution of the system of ordinary differential equations have if there is to be oscillatory behavior of the physical system which is described by the periodic solution? Examples 6.2 and 6.3 show that if small disturbances occur which are not included in the description of the physical system given by the differential equations, then it is not sufficient to find a periodic solution or even a periodic solution which is a limit cycle. Example 6.1 suggests that the only periodic solutions of the differential equations which will be reflected in the behavior of the physical system are those periodic solutions which have some fairly strong asymptotic stability properties. Roughly speaking, if the physical system is "moving along the periodic solution," i.e. if the physical system is described by the periodic solution, then if the physical system is "disturbed" so that it is "kicked onto" another solution of the differential equation, then as the physical system "moves along" the new solution, we require that it approach the original periodic solution.

That is, if we are looking for mathematical descriptions of oscillations in systems in which small disturbances occur (and biological and chemical systems are presumably systems of this kind), we must look for periodic solutions with some kind of asymptotic stability properties, not simply periodic solutions.

The Poincaré-Bendixson Theorem, however deep and interesting it is, does not help in this search because it tells us only that under certain circumstances there exists a periodic solution which may be a limit cycle.

Rather than considering the question of whether there exists a periodic solution with some kind of asymptotic stability properties, we will consider a more general question: what is the behavior of bounded solutions which have asymptotic stability properties? We will prove a striking result for n-dimensional systems due to G. Sell [1966]: the Ω-limit set of a bounded phase asymptotically stable solution is a phase asymptotically stable periodic solution (which may of course be an equilibrium point).

We prove first that the Ω-limit set of a bounded uniformly stable solution contains the orbit of an almost periodic uniformly stable solution.

We consider the n-dimensional autonomous system

$$x' = f(x) \tag{18}$$

where f is a continuous n-vector function with domain U, an open set in R^n such that f satisfies a local Lipschitz condition at each point of U.

Definition. Solution $x(t)$ of (18) is *bounded* if there exist numbers t_0 and M such that for all $t \geq t_0$ solution $x(t)$ is defined and

$$|x(t)| < M$$

Definition. Let $h(t)$ be a real-valued continuous function from the real t-axis into R^n. Then $h(t)$ is *almost periodic* if and only if: given $\varepsilon > 0$ then there exists a number $L(\varepsilon)$ such that in every open interval $(t_0, t_0 + L(\varepsilon))$, where t_0 is a real number, there is a number T_0 such that for all real t,

$$|h(t + T_0) - h(t)| < \varepsilon$$

The set of numbers $\{T_0\}$ is called a *relatively dense set*. Note that a periodic function $p(t)$ is a trivial example of an almost periodic function: if T is the period of $p(t)$, let $L(\varepsilon) = T + \delta$ where $\delta > 0$. Then if t_0 is any real number, there is a number of the form mT, where m is an integer, in the interval $(t_0, t_0 + T + \delta)$ and for all t,

$$|p(t + mT) - p(t)| = 0$$

Sell-Deysach Theorem. *If $x(t)$ is a bounded uniformly stable solution of (18), then there is an almost periodic uniformly stable solution $\bar{x}(t)$ of (18) such that $O[\bar{x}(t)] \subset \Omega[x(t)]$.*

Proof. We divide the proof into three lemmas. The first lemma, which is the main result, is a kind of n-dimensional analog of the Poincaré-Bendixson Theorem.

Definition. A solution $x(t)$ of (18) defined for all real t is *recurrent* if, given $\varepsilon > 0$, then there exists a positive number $T(\varepsilon)$ such that if t_1 and t_2 are real numbers, then there exists $t_3 \in (t_1, t_1 + T(\varepsilon))$ such that

$$|x(t_2) - x(t_3)| < \varepsilon$$

That is, if t_1 is any fixed real number, every point in $0[x(t)]$ is within ε of the curve

$$[x(t)/t \in (t_1, t_1 + T(\varepsilon))]$$

Lemma 6.6 *If $x(t)$ is a bounded solution of (18), there is a recurrent solution $\bar{x}(t)$ of (18) such that $O[\bar{x}(t)] \subset \Omega[x(t)]$*

Proof. By Theorem 3.4, the set $\Omega[x(t)]$ is a nonempty invariant, closed, bounded (and therefore compact) set. Hence, by Theorem 3.5, the set $\Omega[x(t)]$ contains a minimal set.

The proof of Lemma 6.6 follows from:

Birkhoff's Theorem. *Every orbit of a compact minimal set M is recurrent.*

Proof of Birkhoff's Theorem. Let $x(t)$ be a solution such that $O[x(t)] \subset M$, and suppose $x(t)$ is not recurrent. Then there exists a number $r > 0$, a sequence $\{x(\bar{t}_\nu)\}$ and a sequence of pairs $\{t_\nu, T_\nu\}$ such that $T_\nu \to \infty$ and such that if

$$D_\nu = \{x(t)/t \in (t_\nu - T_\nu, t_\nu + T_\nu)\}$$

then for each ν

$$\underset{p \in D_\nu}{\mathrm{g\ell b}} \ |x(\bar{t}_\nu) - p| \geq r$$

By taking subsequences if necessary, let

$$\bar{x} = \lim_{\nu \to \infty} x(\bar{t}_\nu)$$

and

$$\bar{u} = \lim_{\nu \to \infty} x(t_\nu)$$

Let $u(t)$ be a solution of (18) such that

$$u(0) = \bar{u}$$

Let $T > 0$. Since the solution depends continuously on the initial value (Corollary 1.1 to Existence Theorem 1.1), there exists a positive number $\delta = \delta(r/3, T)$ such that if $v(t)$ is a solution of (18) and

$$|u(0) - v(0)| < \delta$$

then for all $t \in [-T, T]$,

$$|u(t) - v(t)| < \frac{r}{3}$$

Let ν be such that

$$T_\nu > T$$
$$|\bar{u} - x(t_\nu)| < \delta$$

and

$$|\bar{x} - x(\bar{t}_\nu)| < \frac{r}{3}$$

Then for any fixed $t \in (-T, T)$:

$$|x(\bar{t}_\nu) - x(t_\nu + t)| \geq r$$

Hence, for $t \in (-T, T)$,

$$\begin{aligned}
|u(t) - \bar{x}| &\geq |x(\bar{t}_\nu) - x(t_\nu + t)| \\
&\quad - |u(t) - x(t_\nu + t)| \\
&\quad - |x(\bar{t}_\nu) - \bar{x}| \\
&\geq \frac{r}{3}
\end{aligned}$$

Since T was chosen arbitrarily, it follows that for all real t,

$$|u(t) - \bar{x}| \geq \frac{r}{3} \tag{19}$$

Since M is closed,

$$\bar{x} \in M$$

and

$$\bar{u} \in M$$

Since M is invariant,

$$O[u(t)] \subset M$$

By (19), $\overline{O[u(t)]}$ is a proper closed subset of M. Since $\overline{O[u(t)]}$ is invariant, we have a contradiction to the hypothesis that M is minimal. This completes the proof of Birkhoff's Theorem and hence the proof of Lemma 6.6 □

Lemma 6.7 *If $x(t)$ is a bounded uniformly stable solution of (18) and if $\tilde{x}(t)$ is a solution of (18) such that $O[\tilde{x}(t)] \subset \Omega[x(t)]$, then $\tilde{x}(t)$ is uniformly stable.*

Proof. Since $x(t)$ is bounded, then $\Omega[x(t)]$ is bounded. Hence, since $\Omega[x(t)]$ is invariant, then $\tilde{x}(t)$ is defined for all real t. Let $t_0 \geq K$, where K occurs in the definition of uniform stability. Since $O[\tilde{x}(t)] \subset \Omega[x(t)]$ there exists $\tau \geq K$ such that

$$|x(\tau) - \tilde{x}(t_0)| < \frac{1}{2} \left(\delta \left[\min \left(\frac{\delta\left(\frac{\varepsilon}{2}\right)}{2}, \frac{\varepsilon}{2} \right) \right] \right) \tag{20}$$

where $\varepsilon > 0$ is given and $\delta(\varepsilon)$ is the function in the definition of uniform stability. Suppose that $u(t)$ is a solution of (18) and there are numbers t_1, t_2 such that $t_2 \geq K$ and

$$|u(t_1) - \tilde{x}(t_2)| < \frac{\delta\left(\frac{\varepsilon}{2}\right)}{2} \tag{21}$$

From (20) and the fact that $x(t)$ is uniformly stable, if $t \geq 0$,

$$|x(t+\tau) - \tilde{x}(t+t_0)| < \min\left(\frac{\delta\left(\frac{\varepsilon}{2}\right)}{2}, \frac{\varepsilon}{2}\right) \tag{22}$$

Assume $t_0 = t_2$. Then it follows from (20) and (21) that

$$|u(t_1) - x(\tau)| < \delta\left(\frac{\varepsilon}{2}\right) \tag{23}$$

Hence, if $t \geq 0$,

$$|u(t+t_1) - x(t+\tau)| < \frac{\varepsilon}{2} \tag{24}$$

Adding (22) and (24), we obtain: if $t \geq 0$

$$|u(t+t_1) - \tilde{x}(t+t_0)| < \varepsilon \tag{25}$$

But $t_0 = t_2$. So (25) is the desired result. This completes the proof of Lemma 6.7.   □

Lemma 6.8 *If $x(t)$ is a recurrent uniformly stable solution of (18), then $x(t)$ is almost periodic.*

Proof. Let K be the constant which appears in the uniform stability condition on $x(t)$. Then the function $y(t) = x(t - K)$ is a recurrent uniformly stable solution of (18) and the constant K in the uniform stability condition on $y(t)$ is zero. We show that $y(t)$ is almost periodic. First, given $\varepsilon > 0$, then by the uniform stability of $y(t)$, there exists a $\delta = \delta\left(\frac{\varepsilon}{2}\right)$. Also, since $y(t)$ is recurrent, there exists $T(\delta) > 0$ such that in every interval of length $T(\delta)$ there is a number T_ν such that

$$|y(T_\nu) - y(0)| < \frac{\delta}{2} \tag{26}$$

(The set $\{T_\nu\}$ is relatively dense.) By the uniform stability of $y(t)$, it follows that if $t \geq 0$,

$$|y(T_\nu + t) - y(t)| < \varepsilon$$

To complete the proof, we need to show that this last inequality also holds for all $t < 0$. For this, we proceed as follows. Let T_ν be fixed. By the continuity in the initial condition of solutions of (18) (Corollary 1.1), there exists $d > 0$ such that if $w(t)$ is a solution of (18) and

$$|y(0) - w(0)| < d$$

then

$$|y(T_\nu) - w(T_\nu)| < \frac{\delta}{2}$$

If t is a given real number, then by the recurrence of $y(t)$, if $A > 0$, there is a real number t_1 such that $t_1 < -A$ and $t_1 < t$ and such that

$$|y(0) - y(t_1)| < \min[d, \delta] \tag{27}$$

Hence by the definition of d,

$$|y(T_\nu) - y(T_\nu + t_1)| < \frac{\delta}{2} \tag{28}$$

From (26) and (28), we have:

$$|y(0) - y(T_\nu + t_1)| < \delta$$

and hence by the uniform stability of $y(t)$, if $t \geq 0$,

$$|y(t) - y(t + T_\nu + t_1)| < \frac{\varepsilon}{2} \tag{29}$$

Since $y(t)$ is uniformly stable with $K = 0$, it follows from (27) that if $t \geq 0$

$$|y(t) - y(t + t_1)| < \frac{\varepsilon}{2} \tag{30}$$

From (29) and (30), it follows that if $t \geq 0$,

$$|y(t + t_1) - y(t + T_\nu + t_1)| < \varepsilon$$

Since t_1 may be chosen so that $t_1 < -A$ where A is a given positive number, this completes the proof of Lemma 6.8, and hence the proof of the Sell-Deysach Theorem.   □

Sell's Theorem. *If $x(t)$ is a bounded phase asymptotically stable solution of (18), then there is a phase asymptotically stable periodic solution $y(t)$ of (18) such that $O[y(t)] = \Omega[x(t)]$.*

Proof. By the Sell-Deysach Theorem, there is an almost periodic uniformly stable solution $\bar{x}(t)$ of (18) such that

$$O[\bar{x}(t)] \subset \Omega[x(t)]$$

Next we need:

Lemma 6.9 *If $x(t)$ is a bounded phase asymptotically stable solution of (18) and if a solution $y(t)$ of (18) is such that*

$$O[y(t)] \subset \Omega[x(t)]$$

then $y(t)$ is phase asymptotically stable.

Proof. From Lemma 6.7, it is sufficient to show that there exists a $\delta > 0$ such that if t_1, t_2 are real numbers such that $t_2 \geq K$ and if a solution $u(t)$ of (18) is such that

$$|u(t_1) - y(t_2)| < \delta$$

then there exists a number t_3 such that

$$\lim_{t \to \infty} |u(t) - y(t_3 + t)| = 0$$

Given $\varepsilon > 0$, let $\delta(\varepsilon)$ be the $\delta(\varepsilon)$ given by the uniform stability of solution $x(t)$. Suppose that there exist numbers t_1, t_2 such that

$$|u(t_1) - y(t_2)| < \frac{\delta(\varepsilon)}{2} \tag{31}$$

Since $O[y(t)] \subset \Omega[x(t)]$, there exists $t_3 \geq K$ such that

$$|x(t_3) - y(t_2)| < \frac{\delta(\varepsilon)}{2} \qquad (32)$$

From (31) and (32), it follows that

$$|u(t_1) - x(t_3)| < \delta(\varepsilon) \qquad (33)$$

and since $x(t)$ is phase asymptotically stable, it follows from (33) that there exists t_4 such that

$$\lim_{t \to \infty} |u(t) - x(t_4 + t)| = 0 \qquad (34)$$

and it follows from (32) that there exists t_5 such that

$$\lim_{t \to \infty} |y(t) - x(t_5 + t)| = 0 \qquad (35)$$

Let

$$\tau = t - (t_4 - t_5)$$

Then (35) becomes

$$\lim_{\tau \to \infty} |y[\tau + (t_4 - t_5)] - x[t_5 + \tau + (t_4 - t_5)]| = 0$$

or

$$\lim_{\tau \to \infty} |y[\tau + (t_4 - t_5)] - x(\tau + t_4)| = 0 \qquad (36)$$

From (34) and (36), we obtain:

$$\lim_{\tau \to \infty} |u(t) - y[t + (t_4 - t_5)]| = 0$$

This completes the proof of Lemma 6.9. $\qquad \square$

By Lemma 6.9, the almost periodic solution $\bar{x}(t)$ is phase asymptotically stable. Since $\bar{x}(t)$ is almost periodic, $O[\bar{x}(t)] \subset \Omega[\bar{x}(t)]$. Suppose $u(t)$ is a recurrent solution of (18) such that

$$O[u(t)] \subset \Omega[\bar{x}(t)]$$

By Lemma 6.9, solution $u(t)$ is phase asymptotically stable and by Lemma 6.8, solution $u(t)$ is almost periodic. Since

$$O[u(t)] \subset \Omega[\bar{x}(t)]$$

and $u(t)$ is phase asymptotically stable, there exists a number t_1 such that

$$\lim_{t \to \infty} |u(t + t_1) - \bar{x}(t)| = 0 \tag{37}$$

Since $u(t)$ is almost periodic, solution $u(t + t_1)$ is almost periodic.

Lemma 6.10 *If f and g are almost periodic functions and if c is a real number, then $cf + g$ is almost periodic.*

Proof. See, for example, Besicovitch [1954], p. 4.

By Lemma 6.10, the function $u(t + t_1) - \bar{x}(t)$ is almost periodic and hence by (37), for all t, $u(t + t_1) = \bar{x}(t)$. Thus the only recurrent solution $u(t)$ such that

$$O[u(t)] \subset \Omega[\bar{x}(t)]$$

is $\bar{x}(t)$ or $\bar{x}(t + k)$ where k is a constant. Since $\Omega[\bar{x}(t)]$ is a nonempty invariant compact set, then by Theorem 3.5 it contains a compact minimal set M. By Birkhoff's Theorem, every orbit in M is the orbit of a recurrent solution. Hence

$$M = O[\bar{x}(t)]$$

Since a minimal set is closed, then

$$O[\bar{x}(t)] = \overline{O[\bar{x}(t)]}$$

Next we use:

Lemma 6.11 *If $u(t)$ is a recurrent solution of (18) such that $u(t)$ is not periodic, then*

$$\overline{O[u(t)]} - O[u(t)] = \Omega[u(t)]$$

Before proving Lemma 6.11, we show how to complete the proof of Sell's Theorem. Since

$$O[\bar{x}(t)] = \overline{O[\bar{x}(t)]}$$

then by Lemma 6.11, $\bar{x}(t)$ is periodic. Also $O[\bar{x}(t)] \subset \Omega[x(t)]$. Since $\bar{x}(t)$ is phase asymptotically stable, it follows that

$$\Omega[x(t)] \subset O[\bar{x}(t)]$$

This completes the proof of Sell's Theorem.

Proof of Lemma 6.11. First we show that

$$\overline{\overline{O[u(t)]} - O[u(t)]} \subset \Omega[u(t)]$$

Since $\Omega[u(t)]$ is closed, it is sufficient to show that

$$\overline{O[u(t)]} - O[u(t)] \subset \Omega[u(t)]$$

Suppose

$$p \in \overline{O[u(t)]} - O[u(t)]$$

Then there is a sequence $\{t_n\}$ such that

$$\lim_{n \to \infty} u(t_n) = p$$

Also $\lim_{n \to \infty} t_n = \infty$ because otherwise p would be a point in $O[u(t)]$. Hence $p \in \Omega[u(t)]$.

In order to prove that

$$\Omega[u(t)] \subset \overline{\overline{O[u(t)]} - O[u(t)]}$$

it is sufficient to prove that

$$O[u(t)] \subset \overline{\overline{O[u(t)]} - O[u(t)]}$$

because then, since

$$\overline{O[u(t)] - O[u(t)]}$$

is closed, it follows that

$$\Omega[u(t)] \subset \overline{O[u(t)] - O[u(t)]}$$

Let $\bar{u} \in O[u(t)]$. It is sufficient to show that, given $\varepsilon > 0$, there exists $\bar{u}^0 \in \overline{B_\varepsilon(\bar{u})}$, where

$$B_\varepsilon(\bar{u}) = \{x \in R^n / |x - \bar{u}| < \varepsilon\}$$

such that $\bar{u}^0 \in \overline{O[u(t)]} - O[u(t)]$. Assume that $u(t)$ is such that $u(0) = \bar{u}$. (If $u(t_0) = \bar{u}$ where $t_0 \neq 0$, just let $u(t)$ denote the solution $v(t) = u(t + t_0)$.) Since $u(t)$ is recurrent, there exists a monotonic strictly increasing sequence $\{t_n\}$ such that $t_n > 0$ for all n, $\lim_{n \to \infty} t_n = +\infty$ and $\lim_{n \to \infty} u(t_n) = \bar{u}$. Choose $\tau_1 > t_1$ such that

$$v_1 = u(\tau_1) \in B_\varepsilon(\bar{u})$$

Then since $\tau_1 > t_1$ and $u(t)$ is not periodic,

$$v_1 \notin \{u(t)/t \in [-t_1, t_1]\}$$

and

$$\delta_1 = \mathop{\mathrm{glb}}_{t \in [-t_1, t_1]} |v_1 - u(t)| > 0$$

Let

$$\varepsilon_1 = \min\left[\frac{\varepsilon}{2}, \varepsilon - |\bar{u} - v_1|, \frac{1}{2}\delta_1\right]$$

Then if

$$B_{\varepsilon_1}(v_1) = \{x \in R^n / |x - v_1| < \varepsilon_1\}$$

we have

$$B_{\varepsilon_1}(v_1) \subset B_\varepsilon(\bar{u}) \tag{38}$$

and

$$\overline{B_{\varepsilon_1}(v_1)} \cap \{u(t)/t \in [-t_1, t_1]\} = \phi$$

Now assume $v_{n-1} = u(\tau_{n-1})$ and ε_{n-1} have been defined. Since $u(t)$ is recurrent, we can choose $\tau_n > t_n$ such that

$$v_n = u(\tau_n) \in B_{\varepsilon_{n-1}}(v_{n-1}) \tag{39}$$

Let

$$\delta_n = \operatorname*{glb}_{t \in [-t_n, t_n]} |v_n - u(t)|$$

and define

$$\varepsilon_n = \min\left[\frac{\varepsilon_{n-1}}{2}, \varepsilon_{n-1} - |v_n - v_{n-1}|, \frac{1}{2}\delta_n\right]$$

Then from the definition of ε_n, it follows that

$$B_{\varepsilon_n}(v_n) \subset B_{\varepsilon_{n-1}}(v_{n-1}) \tag{40}$$

Also from the definition of δ_n

$$\overline{B_{\varepsilon_n}(v_n)} \cap \{u(t)/t \in [-t_n, t_n]\} = \phi \tag{41}$$

The sequence $\{v_n\}$ is such that

$$|v_n - v_{n-1}| < \varepsilon_{n-1} \le \frac{\varepsilon}{2^{n-1}}$$

Thus $\{v_n\}$ is a Cauchy sequence and converges to a point v. Since $\{v_n\} \subset O[u(t)]$, then $v \in \overline{O[u(t)]}$ and since by (38) and (40), $|v_n - \bar{u}| < \varepsilon$ for all n, then $|v - \bar{u}| \le \varepsilon$.

To complete the proof of Lemma 6.11, we show that $v \notin O[u(t)]$. Suppose there exists τ such that

$$v = u(\tau)$$

Take $t_n > |\tau|$. Then

$$v \in \{u(t)/t \in [-t_n, t_n]\} \tag{42}$$

But since, by (40)

$$v \in \overline{B_{\epsilon_n}(v_n)}$$

and, by (41),

$$\overline{B_{\epsilon_n}(v_n)} \cap \{u(t)/t \in [-t_n, t_n]\} = \phi$$

then

$$v \notin \{u(t)/t \in [-t_n, t_n]\}$$

which contradicts (42).   □

Periodic Solutions for Nonautonomous Systems

So far we have looked only at the problem of periodic solutions for autonomous systems. One can ask a parallel question for nonautonomous systems: given the system

$$x' = f(t, x)$$

where f has period T as a function of t, i.e. for all x and all t,

$$f(t + T, x) = f(t, x)$$

then does the system have a solution of period T?

Notice that this question differs from the question about autonomous systems in that it is more limited. We are given a period T and we seek a solution with that period. In studying the autonomous systems we must search for a periodic solution but we have no idea what the value of the period should be. This might lead one to suspect that the question about periodic solutions for nonautonomous equations would be easier to answer. This is, in fact, the case. We will show now that application of the Brouwer Fixed Point Theorem leads easily to the existence of periodic solutions in rather general circumstances. For cases in which the Fixed Point Theorem cannot be applied, the more general theory of topological degree is sometimes applicable.

First we formulate our problem precisely. We consider the n-dimensional equation

$$x' = f(t, x) \tag{43}$$

where f has domain $R \times R^n$ and f has continuous first derivatives in all variables at each point of $R \times R^n$. We assume that f as a function of t has period T, i.e. for each $(t, x) \in R \times R^n$,

$$f(t + T, x) = f(t, x)$$

and we ask the question: does equation (43) have a solution $x(t)$ of period T? To answer this question, we prove first a very simple lemma which is often useful in the study of periodic solutions of nonautonomous systems.

Lemma 6.12 *A solution $x(t)$ of (43) has period T if and only if*

$$x(T) - x(0) = 0 \tag{44}$$

Proof. If $x(t)$ has period T, equation (44) obviously holds. Suppose (44) is true. The function $y(t) = x(t + T)$ is also a solution of (43) because, using the periodicity of f as a function of t, we have:

$$\frac{dy}{dt} = \frac{d}{dt} x(t + T) = f[t + T, x(t + T)] = f[t, x(t + T)] = f[t, y(t)]$$

But by (44),

$$y(0) = x(T) = x(0)$$

Hence by the uniqueness of solution, it follows that for all t

$$x(t+T) = y(t) = x(t)$$

This completes the proof of Lemma 6.12   □

Now we assume that there is a bounded open set U in R^n such that $\bar{U}$ is a homeomorphism of a closed ball and if $x(t,c)$ is the solution of (43) such that

$$x(0,c) = c \in \bar{U}$$

then

$$x(T,c) \in \bar{U}$$

That is, the mapping M defined by

$$M : c \to x(T,c)$$

is a mapping which takes $\bar{U}$ into $\bar{U}$. It follows from Corollary 1.1 in Chapter 1 that M is continuous. Hence by the Brouwer Fixed Point Theorem, it follows that M has a fixed point in $\bar{U}$, i.e. there exists $c_0 \in \bar{U}$ such that

$$x(0,c_0) = c_0 = x(T,c_0)$$

Hence by Lemma 6.12, $x(t,c_0)$ is a solution of period T of equation (43). Summarizing this discussion, we obtain

Theorem 6.1 *Given the n-dimensional system*

$$x' = f(t,x) \tag{45}$$

where f has domain $R \times R^n$ and f has continuous first derivatives in all variables at each point of $R \times R^n$ and there exists a positive number T such that for all $(t,x) \in R \times R^n$

$$f(t+T,x) = f(t,x)$$

suppose there exists a bounded open set $U \subset R^n$ such that U is a homeomorphism of a closed ball and such that if $x(t,c)$ denotes the solution of (45) *with*

$$x(0,c) = c$$

then if $c \in \bar{U}$, it follows that $x(T,c) \in \bar{U}$. Then equation (45) *has a solution $x(t,c_0)$, where $c_0 \in \bar{U}$, such that $x(t,c_0)$ has period T.*

We add a few remarks about the significance of the conditions in this theorem:

1. From the point of view of pure mathematics, the hypothesis on the bounded open set U seems very strong. However, from the point of view of some applications, it is a rather natural condition, as has already been pointed out earlier for the autonomous case.

2. The periodic solution $x(t,c_0)$ may be trivial, i.e. it may happen that $x(t,c_0)$ is a constant solution. This can happen only if for all $t \in [0,T]$,

$$f(t,c_0) = 0 \tag{46}$$

Thus if the theorem is applied and one wishes to conclude that there exists a nontrivial periodic solution, it is sufficient to show that there is no point $c_0 \in \bar{U}$ such that (46) holds for all $t \in [0,T]$.

3. The theorem gives no information about the number of periodic solutions or the stability properties of the periodic solutions. There seems little hope of obtaining such information unless stronger hypotheses are imposed. In the next chapter we will study the problem of periodic solutions of quasilinear nonautonomous systems. In this case, by using more refined topological methods, we will be able to get an estimate on the number of periodic solutions and some information about stability.

4. More refined topological methods, especially use of topological degree, can be used in the study of equation (45). The most notable work is due to Gomory [1956]. For extensive references to other work, see Mawhin and Rouché [1973].

Exercises for Chapter 6

1. Prove that if (x_0, y_0) is an equilibrium point of

$$x' = P(x, y)$$
$$y' = Q(x, y)$$

and if the eigenvalues of the matrix

$$\begin{bmatrix} P_x(x_0, y_0) & P_y(x_0, y_0) \\ Q_x(x_0, y_0) & Q_y(x_0, y_0) \end{bmatrix}$$

both have positive real parts, then (x_0, y_0) is contained in the interior of a simple closed curve such that no solution enters that interior. (Hint: make the transformation of variables $t \to -\tau$ and show that (x_0, y_0) is an asymptotically stable equilibrium point of the resulting system.)

2. Show that $r = 1$ is the orbit of a phase asymptotically stable solution of the system

$$x' == y + \frac{x}{r}(1 - r^2) \tag{1}$$

$$y' = x + \frac{y}{r}(1 - r^2) \tag{2}$$

3. Prove that the function

$$\alpha(r) = -\exp\left(\frac{-1}{(r-1)^2}\right) \qquad r > 1$$

$$\alpha(r) = 0 \qquad\qquad\qquad 0 < r \le 1$$

is a monotonic nonincreasing differentiable function for $r > 0$.

4. A simple way to describe an example which shows that the Poincaré-Bendixson Theorem is not valid for $n > 2$ is to specify a vector field. Let the set

$$\{(u, v)/0 \le u \le 1, 0 \le v \le 1\}$$

in which the points $(0, v)$ and $(1, v)$ are identified and $(u, 0)$ and $(u, 1)$ are identified, be used to describe a torus embedded in R^3. On the torus, specify the vector field by

$$\dot{u} = 1$$
$$\dot{v} = k$$

where k is an irrational number. Extend the vector field to R^3 by any continuous extension. The resulting vector field has no singularities on the torus. Since k is irrational, the solutions of the corresponding differential equation whose orbits pass through a point on the torus are such that their orbits are contained in the torus and are not closed curves. Each such orbit shows that the Poincaré-Bendixson Theorem cannot hold if $n = 3$.

5. Prove the following generalization of Theorem 4.5.
   Theorem. *If $x(t)$ is a nontrivial almost periodic solution of the autonomous system*

   $$x' = f(x)$$

   then $x(t)$ is not asymptotically stable.

   (Hint: use the definition of almost periodic and construct a proof parallel to the proof of Theorem 4.5.)

Chapter 7

Bifurcation and Branching of Periodic Solutions

Introduction

A classical problem in the study of periodic solutions is the following: given the system

$$x' = F(t, x, \varepsilon) \qquad \text{(E)}$$

where ε is a parameter, suppose that for a fixed value of the parameter, say $\varepsilon = 0$, system (E) has a periodic solution. (We include the possibility that the periodic solution is trivial, i.e. constant.) Then does (E) have a periodic solution for small $|\varepsilon|$ and is this periodic solution "near" the given periodic solution?

For mathematicians accustomed to abstract global thinking, this may seem a small uninteresting question. One is inclined also to suspect that the problem admits an easy solution. Actually, the problem is very difficult and, despite a large number of studies over a long period of time, it is still a subject of research.

It is a very important problem for two reasons. First, it arises continually in applications of differential equations: in celestial mechanics, electrical engineering, control theory and biology. From the point of view of pure mathematics, the problem is important because it can be regarded as the prototype of a certain nonlinear question which arises for many types of functional equations.

In Chapter 6, we saw that the solution of a problem about the existence of periodic solutions took very different form in the two cases of autonomous and nonautonomous equations. The same is true of the problem we study in this chapter. If (E) is nonautonomous, the appearance of periodic solutions for small $|\varepsilon|$ is called the branching of periodic solutions. If (E) is autonomous, it is called the bifurcation of periodic solutions. (There is no logical reason for this terminology. It has just gradually come into common usage.) Since quite different approaches are used to study branching and bifurcation, we will deal with them separately. The basic mathematical technique employed is the implicit function theorem which has been applied in branching and bifurcation problems, since the time of Poincaré. We will obtain further results on existence and stability of periodic solutions by using, in addition, topological degree theory. In certain parts of our discussion we use extensively the material on topological degree discussed in the Appendix. The results obtained include and extend those of Friedrichs [1965], Malkin [1959] and Coddington and Levinson [1955].

A. Branching of Periodic Solutions

An Example

Branching of periodic solutions is essentially a nonlinear phenomenon, and we begin by describing a simple example which suggests this. The example shows, in addition, the importance of including nonlinear terms in the description of some physical systems.

The example we will describe is taken from mechanics and is discussed in full detail in Boyce and DiPrima [1986]. For small values of the displacement u, a spring-mass system with no damping and a periodic external force can be described by the linear differential equation

$$mu'' + ku = \varepsilon \cos \omega t \qquad (1)^*$$

where m is the mass and k is the spring constant and the periodic external force is described by $\varepsilon \cos \omega t$ where ε and ω are nonzero constants. Now suppose that

$$\sqrt{\frac{k}{m}} = \omega$$

Then (1)* may be written as:

$$u'' + \omega^2 u = \frac{\varepsilon}{m} \cos \omega t \qquad (2)^*$$

By elementary methods, it is easy to show that the general solution of (2)* is

$$u(t) = c_1 \cos \omega t + c_2 \sin \omega t + \frac{\varepsilon}{2m\omega} t \sin \omega t \qquad (3)^*$$

where c_1, c_2 are arbitrary constants. The expression

$$c_1 \cos \omega t + c_2 \sin \omega t$$

is the general solution of the homogeneous equation

$$u'' + \omega^2 u = 0$$

and the expression

$$\frac{\varepsilon}{2m\omega} t \sin \omega t$$

is a particular solution of the inhomogeneous equation (2)*. Thus for $\varepsilon = 0$, equation (2)* has a periodic solution (in fact all the solutions of (2)* are periodic), but if $\varepsilon \neq 0$, no solution of (2)* is periodic (and thus branching of periodic solutions does not occur). Looking at the general solution (3)* of equation (2)*, we see that if $\varepsilon \neq 0$, each solution of (2)* is oscillatory in the sense that as t increases, the solution assumes alternately positive and negative values, but the amplitudes of the oscillations increase without bound. This is sometimes called a "resonance catastrophe" (not to be confused with R. Thom's catastrophe theory which is something entirely different) or an infinity catastrophe. Actually, as t increases and the amplitudes of the oscillations of $u(t)$ increase, equation (1)* is no longer a valid description of the oscillations of the spring-mass system. This follows from the derivation of equation (1)* in which it is assumed that u remains small so that the higher-order terms in u can be dropped. As $|u(t)|$ increases, the nonlinear effects begin to play a more important role and the mathematical description or model of the spring-mass system becomes a nonlinear differential equation. Since in physical reality the amplitudes of the oscillations of the spring-mass system remain bounded, it is reasonable to ask if branching does occur in this nonlinear model. We will see that branching often does occur if the differential equation is nonlinear.

Existence of Periodic Solutions

We proceed now to a general study of the branching problem. We consider the n-dimensional system

$$u' = f(t, u, \varepsilon) \tag{1}$$

where the n-vector function f has continuous third derivatives at each point

$$(t, u, \varepsilon) \in R \times R^n \times I$$

where I is an open interval on the real line with midpoint zero and f has period $T(\varepsilon)$ in t where $T(\varepsilon)$ is a positive-valued differentiable function with domain I. We assume that for $\varepsilon = 0$, equation (1) has a solution $u(t)$ of period $T(0)$ and study the following problem.

Problem 1. If $|\varepsilon|$ is sufficiently small, does equation (1) have a solution $u(t, \varepsilon)$ of period $T(\varepsilon)$ such that for each real t,

$$\lim_{\varepsilon \to 0} u(t, \varepsilon) = u(t)$$

First, we simplify the problem by introducing the variable

$$s = \left[\frac{T(0)}{T(\varepsilon)}\right] t$$

and letting

$$g(s, u, \varepsilon) = \frac{T(\varepsilon)}{T(0)} f\left(\frac{T(\varepsilon)}{T(0)} s, u, \varepsilon\right)$$

Then equation (1) becomes

$$\frac{du}{ds} = g(s, u, \varepsilon) \tag{2}$$

and since $g(s, u, \varepsilon)$ has period $T(0)$ in s, Problem 1 becomes:

Problem 2. If $|\varepsilon|$ is sufficiently small, does equation (2) have a solution $u(s, \varepsilon)$ of period $T(0)$ such that for all $s \in [0, T(0)]$,

$$\lim_{\varepsilon \to 0} u(s, \varepsilon) = u(s)$$

In order to stay with conventional notation, we rewrite equation (2) as

$$\frac{du}{dt} = g(t, u, \varepsilon) \tag{3}$$

and we investigate Problem 2 by investigating solutions of equation (3) of the form

$$u(t, \varepsilon) = u(t) + \varepsilon x(t, \varepsilon) \tag{4}$$

where $u(t)$ is the given solution of period $T(0)$. Substituting from (4) into (3) and using Taylor's expansion with a remainder, we obtain

$$\begin{aligned}
u' + \varepsilon x' &= g(t, u + \varepsilon x, \varepsilon) \\
&= g(t, u, 0) + \{g_u[t, u(t), 0]\}\varepsilon x \\
&\quad + \{g_\varepsilon[t, u(t), 0]\}\varepsilon + \varepsilon^2 G[t, x, \varepsilon]
\end{aligned} \tag{5}$$

where the twice differentiable $n \times n$ matrix $g_u[t, u(t), 0]$ has period $T = T(0)$ in t, the n-vector function $g_\varepsilon[t, u(t), 0]$ is a twice differentiable function of t which has period T and the function G has continuous first derivatives in all variables and has period T in t. Since $u(t)$ is a solution of (3) with $\varepsilon = 0$, i.e.

$$u' = g(t, u, 0) \tag{6}$$

then subtracting (6) from (5) and dividing by ε, we obtain:

$$x' = \{g_u[t, u(t), 0]\}x + \varepsilon\mathcal{G}[t, x, \varepsilon] + g_\varepsilon[t, u(t), 0] \tag{7}$$

Since $g_u[t, u(t), 0]$ has period T in t, then according to the Floquet theory there exists a transformation of the dependent variable so that the equation

$$x' = \{g_u[t, u(t), 0]\}x$$

can be reduced to the form

$$x' = Ax$$

where A is a constant matrix in real canonical form. (See Chapter 2.) Thus the problem of studying equation (3) is reduced first to the study of equation (7) and then, by using the Floquet transformation, to the study of the following equation:

$$x' = Ax + \varepsilon F(t, x, \varepsilon) + G(t) \tag{8}$$

where A is a constant matrix in real canonical form, the functions F and G have continuous first derivatives in all variables and F and G have period T in t. Thus Problem 2 can be rephrased as:

Problem 3. If $|\varepsilon|$ is sufficiently small, does equation (8) have solutions of period T?

We make a detailed analysis of Problem 3. The initial steps in the analysis form a classical procedure which was originated by Poincaré and has been widely used ever since. From Existence Theorem 2.1 for Linear Systems (Chapter 2) and the Existence Theorem for Equation with a Parameter (Chapter 1), it follows that if c is a fixed real n-vector and if $|\varepsilon|$ is sufficiently small, there exists a solution $x(t, \varepsilon, c)$ of equation (8) which is defined on an open interval which contains $[0, T]$ and which satisfies the initial condition

$$x(0, \varepsilon, c) = c$$

By the variation of constants formula (Chapter 2), solving equation (8) for $x(t, \varepsilon, c)$ is equivalent to solving the following integral equation for $x(t,\varepsilon,c)$.

$$x(t,\varepsilon,c) = e^{tA}c + e^{tA}\int_0^t e^{-sA}\{\varepsilon F[x(s,\varepsilon,c),s,\varepsilon] + G(s)\}ds \qquad (9)$$

In order to search for solutions $x(t,\varepsilon,c)$ which have period T, we use the following simple but useful lemma.

Lemma 7.1 *A nasc that $x(t,\varepsilon,c)$ have period T is that*

$$x(T,\varepsilon,c) - x(0,\varepsilon,c) = 0 \qquad (10)$$

Proof. The condition is obviously necessary. To show that it is sufficient, define the function $y(t,\varepsilon,c)$ by

$$y(t,\varepsilon,c) = x(T+t,\varepsilon,c)$$

Then $y(t,\varepsilon,c)$ is a solution of equation (8) because

$$y'(t,\varepsilon,c) = x'(T+t,\varepsilon,c)$$
$$= Ax(T+t,\varepsilon,c) + F[T+t,x(T+t,\varepsilon,c),\varepsilon] + G(T+t)$$

Since F and G have period T in t, this equation becomes:

$$y'(t,\varepsilon,c) = Ay(t,\varepsilon,c) + F[t,y(t,\varepsilon,c),\varepsilon] + G(t)$$

Thus $y(t,\varepsilon,c)$ is a solution of equation (8). Also, since

$$y(0,\varepsilon,c) = x(T,\varepsilon,c) = x(0,\varepsilon,c)$$

then by the uniqueness of solution of (8), it follows that for all t,

$$y(t,\varepsilon,c) = x(t,\varepsilon,c)$$

That is,

$$x(T+t,\varepsilon,c) = x(t,\varepsilon,c)$$

for all t. This completes the proof of Lemma 7.1   □

We use equation (10) to search for periodic solutions in this way: since the general solution $x(t, \varepsilon, c)$ of equation (8) is (in some abstract sense if not explicitly) known, then we try to solve equation (10) for c as a function of ε. In order to do this, we first write (10) in as explicit a form as possible and for this purpose, it is convenient to use equation (9). Taking $t = T$ in (9), substituting in (10), and using the equality $x(0, \varepsilon, c) = c$, we obtain:

$$(e^{TA} - I)c + e^{TA} \int_0^T e^{-sA}\{\varepsilon F[s, x(s, \varepsilon, c), \varepsilon] + G(s)\}ds = 0 \quad (11)$$

In summary, to solve Problem 3, it is sufficient to solve equation (11) for c as a function of ε. (Remember that equation (11) is an n-vector equation. We will find it convenient sometimes to regard (11) as a system of n real equations in $\varepsilon, c_1, \ldots, c_n$ where $c_1, \ldots, c_n$ are the components of c.)

First, if $\varepsilon = 0$, equation (11) becomes

$$(e^{TA} - I)c + e^{TA} \int_0^T e^{-sA}G(s)ds = 0$$

Thus, if the matrix $(e^{TA} - I)$ is nonsingular, equation (11) has the initial solution:

$$\varepsilon = 0, \quad c_0 = -(e^{TA} - I)^{-1}e^{TA} \int_0^T e^{-sA}G(s)ds$$

Also, if $e^{TA} - I$ is nonsingular, the Implicit Function Theorem can be applied to solve equation (11) uniquely for c as a function of ε in a neighborhood of this initial solution. Since A is in real canonical form (see Chapter 2), it is easy to see that $e^{TA} - I$ is nonsingular if and only if the eigenvalues of the matrix TA are all nonzero and are all different from $\pm i2n\pi$ $(n = 1, 2, \ldots)$. This last condition is equivalent to the condition that the equation

$$x' = Ax$$

has no nontrivial solutions of period T. (See Exercise 1.) Thus, we have obtained the following classical result (proved by Poincaré):

Theorem 7.1 *If the equation*

$$y' = \{g_u[t, u(t), 0]\}y$$

(i.e. the linear variational system of (3) relative to the given periodic solution $u(t)$) has no nontrivial solutions of period $2n\pi/T$ where $n = 0, 1, 2, \ldots$ (or, equivalently, if matrix A in equation (8) has no eigenvalues of the form $(2n\pi/T)i$ ($n = 0, 1, 2, \ldots$)) then there exist $\eta_1 > 0$, $\eta_2 > 0$ such that for each ε with $|\varepsilon| < \eta_1$, there is a unique vector $c = c(\varepsilon)$, where $c(\varepsilon)$ is a continuous function of ε, such that

$$|c(\varepsilon) - c_0| < \eta_2$$

and such that the solution

$$u(t, \varepsilon) = u(t) + \varepsilon x(t, \varepsilon, c(\varepsilon))$$

of (3) has period T.

(For the relation of Theorem 7.1 to Problem 1, see Exercise 2.)

Next we suppose that $(e^{TA} - I)$ is a singular matrix. This is sometimes called the resonance case. In applications in physics and engineering, the resonance case is often more important than the nonresonance case, i.e. the case considered in Theorem 7.1. In studying the resonance case, we will use a finite-dimensional version of the Lyapunov-Schmidt procedure. Let E_{n-r} denote the null space of $e^{TA} - I$ and let E_r denote the complement in R^n of E_{n-r}, i.e.

$$R^n = E_{n-r} \oplus E_r \quad \text{(direct sum)}$$

Let P_{n-r}, P_r denote the projections of R^n onto E_{n-r} and E_r respectively. Then $P_r^2 = P_r$, $P_{n-r}^2 = P_{n-r}$, and $P_{n-r}P_r = P_rP_{n-r} = 0$. Temporarily, we assume that E_{n-r} is properly contained in R^n, i.e. we assume that E_r is an r-dimensional space where $0 < r < n$. Later we will describe the simplification which occurs if $E_r = 0$, i.e.

$E_{n-r} = R^n$. It will be convenient in the computations to use a real nonsingular matrix H such that

$$H(e^{TA} - I) = P_r$$

Since A is in real canonical form, matrix H can be explicitly computed. (See Exercise 3.) We notice that since $P_{n-r}P_r = 0$ and

$$H(e^{TA} - I) = P_r$$

then, applying P_{n-r} to this last equation, we obtain:

$$P_{n-r}He^{TA} = P_{n-r}H \tag{12}$$

Now we are ready to investigate equation (11) in the resonance case, i.e. the case in which the matrix $e^{TA} - I$ is singular. First we apply matrix H to equation (11) and obtain:

$$P_r c + He^{TA}\int_0^T e^{-sA}\{\varepsilon F[s, x(s, \varepsilon, c), \varepsilon] + G(s)\}ds = 0 \tag{13}$$

Since H is a nonsingular matrix, solving equation (11) for c is equivalent to solving equation (13) for c. Applying P_{n-r} to (13) yields:

$$P_{n-r}He^{TA}\int_0^T e^{-sA}\{\varepsilon F[s, x(s, \varepsilon, c), \varepsilon] + G(s)\}ds = 0 \tag{14}$$

Setting $\varepsilon = 0$ in (14), we obtain:

$$P_{n-r}He^{TA}\int_0^T e^{-sA}G(s)ds = 0$$

and from (12) this equation becomes

$$P_{n-r}H\int_0^T e^{-sA}G(s)ds = 0$$

Thus we have obtained a version of another classical result which we state as the following theorem.

Theorem 7.2 *A necessary condition that Problem 3 can be solved in the resonance case (i.e. that equation* (11) *can be solved for c in terms of ε if $e^{TA} - I$ is a singular matrix) is that*

$$P_{n-r} H \int_0^T e^{-sA} G(s)ds = 0 \qquad (15)$$

Condition (15) really stems the standard orthogonality condition for the solution of nonhomogeneous linear algebraic equations (Theorem 4A in the discussion of Sturm-Liouville theory in Chapter 2). A little later we will see illustrations of its meaning in examples.

In the remainder of the discussion of branching of periodic solutions we assume that the necessary condition (15) *in Theorem 7.2 is satisfied.*

For simplicity in calculations we assume also that

$$P_r H e^{TA} \int_0^T e^{-sA} G(s)ds = 0$$

although the steps we will take can be carried out (with more labor and detail) if this assumption is not made.

Now we proceed to solve equation (13) for c in terms of ε. Applying P_r and P_{n-r} to (13) and using (15) and (12), we obtain:

$$P_r c + P_r H e^{TA} \int_0^T e^{-sA} \{\varepsilon F[s, x(s, \varepsilon, c), \varepsilon] + G(s)\}ds = 0 \qquad (16)$$

$$P_{n-r} H \int_0^T e^{-sA} \varepsilon F[s, x(s, \varepsilon, c), \varepsilon]ds = 0 \qquad (17)$$

Let $P_r c = \xi$, an r-vector, and let $P_{n-r} c = \eta$, an $(n-r)$-vector. Then (16) may be written as:

$$\xi + P_r H e^{TA} \int_0^T e^{-sA} \{\varepsilon F[s, x(s, \varepsilon, \xi \oplus \eta), \varepsilon] + G(s)\} ds = 0 \qquad (18)$$

Equation (18) has the initial solution

$$\varepsilon = 0$$
$$\eta = 0$$
$$\xi = -P_r H e^{TA} \int_0^T e^{-sA} \{G(s)\} ds = 0$$

and the Jacobian of the left-hand side of (18) with respect to ξ at the initial condition is:

$$\det \begin{bmatrix} 1 & \cdot & \cdot & \cdot & 0 \\ & 1 & & & \\ \vdots & & & \ddots & \\ 0 & & & & 1 \end{bmatrix} = 1 \neq 0$$

Hence by the Implicit Function Theorem we can solve uniquely for ξ in terms of η and ε in a neighborhood of the initial solution to obtain:

$$\xi = \xi(\eta, \varepsilon) \tag{19}$$

where $\xi(\eta, \varepsilon)$ is differentiable in η and ε and

$$\xi(\eta, 0) = 0$$

Substituting from (19) into (17), we reduce the problem to that of solving

$$P_{n-r} H \int_0^T e^{-sA} \varepsilon F[s, x(s, \varepsilon, \xi(\eta, \varepsilon) \oplus \eta), \varepsilon] ds = 0 \tag{20}$$

for η in terms of ε. First divide (20) by ε. Then for fixed ε, the left-hand side of (20) may be regarded as a mapping of η, an $(n-r)$-vector, into an $(n-r)$-vector, i.e. a mapping M_ε from $R^{(n-r)}$ into $R^{(n-r)}$. We study the solutions of (20) by investigating the topological degree at zero of M_ε. The mapping M_ε is not given very explicitly because it is defined in terms of the general solution $x(s, \varepsilon, c)$. While existence theorems assure us that $x(s, \varepsilon, c)$ exists and has suitable properties of continuity and differentiability, we have, in general, no idea of the explicit form of $x(s, \varepsilon, c)$. So there is little hope of computing or estimating the topological degree of M_ε if $\varepsilon \neq 0$. However, it is considerably easier to study the mapping M_0. This mapping is given quite explicitly and in many cases, the degree can be computed (later we will describe examples). By the invariance under homotopy of the degree (see the discussion in the Appendix) it follows that if $|\varepsilon|$ is sufficiently small, the degree of M_ε is defined and is equal to the degree of M_0.

Let $c = (c_1, \ldots, c_{n-r}, 0, \ldots, 0)$. Then mapping M_0 is described by

$$M_0 : (c_1, \ldots, c_{n-r}) \rightarrow P_{n-r} H \int_0^T e^{-sA} \{F[s, x(s, 0, c), 0]\} ds \qquad (21)$$

Let $\mathcal{B}$ be a ball with center zero in $R^{(n-r)}$. We consider $\deg[M_0, \mathcal{B}, 0]$. There is no theoretical reason why we need only study the degree relative to a ball $\mathcal{B}$. We could consider the closure of any suitable bounded open set instead of $\mathcal{B}$. But it turns out that it is convenient and sufficient for our purposes to consider a ball $\mathcal{B}$.

We have already computed $P_{n-r} H$ and consequently we could write mapping M_0 more explicitly than is done in (21) above. But a more explicit description of M_0 in the general case offers no particular advantages. Consequently, we wait until consideration of examples to write out mapping M_0 in its explicit form. We summarize the discussion above as a theorem.

Theorem 7.3 *If*

$$\deg[M_0, \mathcal{B}, 0] \neq 0$$

then if $|\varepsilon|$ is sufficiently small, equation (8) has a solution $x(t, \varepsilon)$ of period T and $x(0, \varepsilon) = c$ where $c \in B^n$.

We obtain also information about the number of periodic solutions $x(t, \varepsilon)$. That is, we have the following theorem.

Theorem 7.4 *If*

$$\deg[M_0, \mathcal{B}, 0] \neq 0$$

and if $\delta > 0$, then there exists differentiable function $h(t)$ with the following properties:

(i) *$h(t)$ has period T;*

(ii) *$\max\limits_{t} |h(t)| < \delta$;*

(iii) *if the function $F(t, x, \varepsilon)$ in equation (8) is replaced by $F(t, x, \varepsilon)$ $+h(t)$, then if $|\varepsilon|$ is sufficiently small, the number of solutions of (8) of period T is greater than or equal to $|\deg[M_0, \mathcal{B}, 0]|$;*

(iv) *each periodic solution $x(t, \varepsilon)$ depends continuously on ε.*

Theorem 7.4 says roughly that if function F is varied arbitrarily slightly by adding an arbitrarily small function of t, then the absolute value of the degree is a lower bound for the number of periodic solutions and that the periodic solutions depend continuously on ε. Before giving a detailed proof, we give a brief outline of it. The basic idea of the proof is a straightforward application of Theorem 1 in the Appendix. By adding the function $h(t)$ to F, the mapping M_0 is changed so that $M_0^{-1}(0)$ is a finite set of points, at each of which the Jacobian of M_0 is nonzero. Conclusion (iii) then follows from Theorem 1 in the Appendix. Conclusion (iv) follows from the Implicit Function Theorem. As will be seen in the detailed proof, we have a wide choice for the function $h(t)$. Function $h(t)$ need only satisfy rather simple conditions.

The main complication in giving a detailed proof is the description of the conditions on function $h(t)$. The complication arises from the form of the matrix A. For simplicity, we will describe the conditions on the function $h(t)$ for a specific, fairly typical matrix A. Once the particular description has been given, it is easy to see how to obtain the conditions on function $h(t)$ for other matrices A.

Let us assume that matrix A has the form

$$A = \begin{bmatrix} \begin{bmatrix} 0 & \beta & 1 & 0 \\ -\beta & 0 & 0 & 1 \\ & & 0 & \beta \\ & & -\beta & 0 \end{bmatrix} & & & \\ & \begin{bmatrix} 0 & 1 & \\ & 0 & 1 \\ & & 0 \end{bmatrix} & & \\ & & \begin{bmatrix} \cdot & \\ & \cdot \end{bmatrix} & \\ & & & \begin{bmatrix} & \\ & \end{bmatrix} \end{bmatrix}$$

where $\beta = 2n\pi/T$ and n is a nonzero integer, positive or negative, and all the boxes on the main diagonal except the first two are associated with eigenvalues λ such that the real part of λ is nonzero or, if λ is pure imaginary, $\lambda \neq (2\pi n/T)i$ (where $n = \pm 1, \pm 2, \dots$). Then

$$e^{tA} =$$

$$\begin{bmatrix} \begin{bmatrix} \cos\beta t & \sin\beta t & t\cos\beta t & t\sin\beta t \\ -\sin\beta t & \cos\beta t & -t\sin\beta t & t\cos\beta t \\ & & \cos\beta t & \sin\beta t \\ & & -\sin\beta t & \cos\beta t \end{bmatrix} & & & \\ & \begin{bmatrix} 1 & t & \frac{t^2}{2!} \\ & 1 & t \\ & & 1 \end{bmatrix} & & \\ & & \begin{bmatrix} \cdot & \\ & \cdot \\ & & \cdot \end{bmatrix} & \\ & & & \ddots \\ & & & & \begin{bmatrix} \cdot & \\ & \cdot \end{bmatrix} \end{bmatrix}$$

and

$$
P_{n-r}H = \begin{bmatrix} \begin{bmatrix} 0 & 0 & 1 & 0 \\ 0 & 0 & 0 & 1 \\ 0 & 0 & 0 & 0 \\ 0 & 0 & 0 & 0 \end{bmatrix} & & & \\ & \begin{bmatrix} 0 & 0 & 1 \\ 0 & 0 & 0 \\ 0 & 0 & 0 \end{bmatrix} & & \\ & & \ddots & \end{bmatrix}
$$

where all the remaining boxes on the main diagonal are zero. Hence

$$
P_{n-r}He^{-sA} = \begin{bmatrix} \begin{matrix} 0 & 0 & \cos\beta s & -\sin\beta s \\ 0 & 0 & \sin\beta s & \cos\beta s \\ 0 & 0 & 0 & 0 \\ 0 & 0 & 0 & 0 \end{matrix} & & \\ & \begin{matrix} 0 & 0 & 1 \end{matrix} & \\ & & \ddots \end{bmatrix}
$$

where all omitted entries are zero. We obtain finally:

$$
P_{n-r}H \int_0^T e^{-sA} F[s, x(s, 0, c), 0]ds
$$

$$
= \int_0^T \begin{bmatrix} (\cos\beta s)F_3 & + & (-\sin\beta s)F_4 \\ (\sin\beta s)F_3 & + & (\cos\beta s)F_4 \\ & 0 & \\ & 0 & \\ & F_7 & \\ & 0 & \\ & \vdots & \\ & 0 & \end{bmatrix} ds
$$

A basis for the subspace E_{n-r} is the set of vectors

$$\begin{bmatrix} 1 \\ 0 \\ 0 \\ \cdot \\ \cdot \\ \cdot \\ 0 \end{bmatrix}, \quad \begin{bmatrix} 0 \\ 1 \\ 0 \\ \cdot \\ \cdot \\ \cdot \\ 0 \end{bmatrix}, \quad \begin{bmatrix} 0 \\ 0 \\ 0 \\ 0 \\ 1 \\ 0 \\ \vdots \\ 0 \end{bmatrix}$$

The mapping M_0 can be described by:

$$M_0(c_1, c_2, c_3) \rightarrow \left(\int_0^T \{(\cos \beta s)F_3[s, x(s,0,c),0] + (-\sin s)F_4[\quad]\} \, ds, \right.$$

$$\int_0^T \{(\sin \beta s)F_3[\quad] + (\cos s)F_4[\quad]ds\},$$

$$\left. \int_0^T F_7[\quad]ds \right)$$

Let $h_3(t)$, $h_4(t)$, $h_7(t)$ be differentiable functions of period T such that

$$\int_0^T [(\cos \beta s)h_3(s) + (-\sin \beta s)h_4(s)]ds \neq 0$$

$$\int_0^T [(\sin \beta s)h_3(s) + (\cos \beta s)h_4(s)]ds \neq 0$$

$$\int_0^T h_7(s)ds \neq 0$$

E.g., suppose $h_3(s) = \cos \beta s$, $h_4(s) = -\cos \beta s$, and $h_7(s) = \cos^2 s$. (Note that we have a very wide latitude in our choice of $h_3(s)$, $h_4(s)$

and $h_7(s)$.) If δ_3, δ_4, δ_7 are sufficiently small in absolute value, then

$$\max_t |\delta_3 h_3(t) + \delta_4 h_4(t) + \delta_7 h_7(t)| < \delta$$

Also, if F_3, F_4, F_7 are replaced with $F_3 + \delta_3 h_3$, $F_4 + \delta_4 h_4$ and $F_7 + \delta_7 h_7$ respectively, the mapping M is replaced by $M + D$ where D is the constant 3-vector:

$$\begin{bmatrix} \displaystyle\int_0^T [\delta_3(\cos \beta s)h_3(s) + \delta_4(-\sin \beta s)h_4(s)]ds \\[2em] \displaystyle\int_0^T [\delta_3(\sin \beta s)h_3(s) + \delta_4(\cos \beta s)h_4(s)]ds \\[2em] \displaystyle\int_0^T \delta_7 h_7(s)ds \end{bmatrix}$$

Since $\deg[M_0, \mathcal{B}, 0] \neq 0$, then by Theorem 1 in the Appendix, it follows that for most sufficiently small $|\delta_3|$, $|\delta_4|$, and $|\delta_7|$, the equation

$$M(c_1, c_2, c_3) + D = 0$$

has a finite set of solutions, that there are at least $|\deg[M_0, \mathcal{B}, 0]$ such solutions and also that the Jacobian of M at each solution is nonzero. This completes the proof of Theorem 7.4.   □

Before going further with this analysis, we point out something of the nature of the solutions obtained. If $\deg[M_0, \mathcal{B}, 0] \neq 0$ then it follows that there exist an n-vector function $c(\varepsilon) = (c_1(\varepsilon), \ldots, c_n(\varepsilon))$ such that the solution $x(t, \varepsilon, c(\varepsilon))$ of equation (8) has solution T. In general, the point $(c(0), \ldots, c_n(0))$ is not zero. Thus, the solution $x(t, \varepsilon, c(\varepsilon))$ of equation (8) "branches away" from the point $(c_1(0), \ldots, c_n(0))$. Thus, we have obtained the remarkable fact that periodic solutions "branch away" from only a finite number of points $(c_1, \ldots, c_n)$. Note that since solutions of the original equation (3) are of the form

$$u(t, \varepsilon) = u(t) + \varepsilon x(t, \varepsilon) \tag{4}$$

where $u(t)$ is the given periodic solution, then all the periodic solutions of (3) that are obtained by using Theorem 7.3 or 7.4 must branch away from the periodic solution $u(t)$. Later, we will obtain a generalization of a theorem of Malkin that shows that periodic solutions branch away from only a finite number of the periodic solutions in a family of periodic solutions of the unperturbed equation.

Stability of the Periodic Solutions

Next, we show that the topological degree yields some information about the stability properties of the periodic solutions that have been obtained.

We seek criteria for the asymptotic stability of the periodic solutions obtained in the preceding section. Suppose that (c_0, ε_0) is a solution of the equation (11) which expresses the periodicity condition, i.e. suppose that

$$\tilde{F}(c_0, \varepsilon_0) \stackrel{\text{def}}{=} (e^{TA} - I)c_0 + e^{TA} \int_0^T e^{-sA} \{\varepsilon_0 F[s, x(s, \varepsilon_0, c_0), \varepsilon_0]$$

$$+ G(s)\} ds = 0 \quad (22)$$

Then the solution $x(t, \varepsilon_0, c_0)$ of equation (8) has period T. Since our object is to study the stability properties of the periodic solution $x(t, \varepsilon_0, c_0)$, the first step is to compare $x(t, \varepsilon_0, c_0)$ with another solution, say $x(t, \varepsilon_0, c_1)$, of equation (8) with $\varepsilon = \varepsilon_0$. Substituting $x(t, \varepsilon_0, c_0)$ and $x(t, \varepsilon_0, c_1)$ into equation (8) and subtracting the first equation from the second, we obtain, by using Taylor's Expansion:

$$x'(t, \varepsilon_0, c_1) - x'(t, \varepsilon_0, c_0) = A[x(t, \varepsilon_0, c_1) - x(t, \varepsilon_0, c_0)]$$
$$+ \varepsilon_0\{F[t, x(t, \varepsilon_0, c_1), \varepsilon_0] - F[t, x(t, \varepsilon, c_0), \varepsilon_0]\}$$
$$= A[x(t, \varepsilon_0, c_1) - x(t, \varepsilon_0, c_0)]$$
$$+ \varepsilon_0\{B(t, \varepsilon_0)[x(t, \varepsilon_0, c_1) - x(t, \varepsilon_0, c_0)]$$
$$+ H[x(t, \varepsilon_0, c_1) - x(t, \varepsilon_0, c_0), t, \varepsilon_0]\} \quad (23)$$

where $B(t, \varepsilon_0)$ is the matrix

$$F_x[t, x(t, \varepsilon_0, c_0), \varepsilon_0]$$

the elements of which are functions of t which have period T, and if $\xi \in R^n$,

$$|H(\xi, t, \varepsilon_0)| = o(|\xi|) \tag{24}$$

for all $t \in [0, T]$, i.e.

$$\lim_{|\xi| \to 0} \frac{|H(\xi, t, \varepsilon_0)|}{|\xi|} = 0$$

and this limit is uniform in $t \in [0, T]$.

Let us first suppose that the eigenvalues of matrix A all have negative real parts. Then by Theorem 7.1, equation (8) has a unique periodic solution $x(t, \varepsilon)$ near 0 for each ε such that $|\varepsilon|$ is sufficiently small. Since the eigenvalues of A have negative real parts, then the solution $u(t) \equiv 0$ of

$$u' = Au$$

is asymptotically stable. Hence, if $|\varepsilon_0|$ is sufficiently small, then by Theorem 4.3, the solution $u(t) \equiv 0$ of

$$u' = [A + \varepsilon_0 B(t, \varepsilon_0)]u$$

is asymptotically stable. By condition (24) on H, it follows from the Stability Theorem for Linear Systems and the Corollary to Stability and Instability Theorems for Nonlinear Systems (both in Chapter 4) that 0 is an asymptotically stable solution of

$$u' = [A + \varepsilon_0 B(t, \varepsilon_0)]u + \varepsilon_0 H[u, t, \varepsilon_0]$$

and hence that $x(t, \varepsilon_0, c_0)$ is asymptotically stable. Thus we have proved:

Theorem 7.5 *If the eigenvalues of matrix A all have negative real parts and if $|\varepsilon|$ is sufficiently small, then equation (8) has a unique solution $x(t, \varepsilon)$ of period T near 0 and $x(t, \varepsilon)$ is asymptotically stable.*

Similar arguments yield the following theorem (see Exercise 4).

Theorem 7.6 *If all the eigenvalues of A are different from $(2n\pi/T)i$ $(n = 0, \pm 1, \pm 2, \dots)$ and if A has an eigenvalue with positive real part, then equation (8) has a unique solution $x(t, \varepsilon)$ of period T near zero and $x(t, \varepsilon)$ is unstable.*

Now we turn to stability of periodic solutions in the resonance case. From (24) and the fact that the matrix $B(t, \varepsilon_0)$ has period T, it follows (by the Stability Theorem for Periodic Solutions in Chapter 4) that in order to study the stability of $x(t, \varepsilon_0, c_0)$, it is sufficient to study the characteristic multipliers of the matrix

$$A + \varepsilon_0 B(t, \varepsilon_0)$$

Let $U(t)$ be the fundamental matrix of the equation

$$w' = [A + \varepsilon_0 B(t, \varepsilon_0)]w$$

such that $U(0)$ is the identity matrix. Then the characteristic multipliers to be studied are the eigenvalues of $U(T)$. If we let $u(t) = x(t, \varepsilon_0, c_1) - x(t, \varepsilon_0, c_0)$ and rewrite equation (23) as:

$$u'(t) = Au(t) + \varepsilon_0 B(t, \varepsilon_0)u(t) + \varepsilon_0 H[u(t), t, \varepsilon + 0] \qquad (25)$$

then by the variation of constants formula, we obtain:

$$u(T) = U(T)u(0) + U(T) \int_0^T [U(s)]^{-1}\varepsilon_0 H[u(s), s, \varepsilon_0]ds \qquad (26)$$

If $u(0) = c_1 - c_0 = K$, a constant n-vector, then (26) may be written:

$$u(T) = U(T)K + U(T) \int_0^T [U(s)]^{-1}\varepsilon_0 H[u(s, K), s, \varepsilon_0]ds \qquad (27)$$

where we now denote $u(s)$ by $u(s, K)$ so that we can indicate the initial condition

$$u(0, K) = K$$

By uniqueness of solution, for all s,

$$u(s,0) = 0$$

and therefore

$$u(s,K) = u(s,K) - u(s,0) = \left[\frac{\partial u}{\partial K}(s,0)\right] K + R(K)$$

where

$$|R(K)| = o(|K|) \tag{28}$$

From condition (24) on H and from (28), it follows that the differential at 0 of the mapping

$$\tilde{M}: K \to u(T) = u(T,K)$$

i.e.

$$\tilde{M}: K \to U(T)K + U(T)\int_0^T [U(s)]^{-1}\varepsilon_0 H[u(s,K),s,\varepsilon_0]ds$$

is the matrix $U(T)$. But we have also

$$u(T) = x(T,\varepsilon_0,c_1) - x(T,\varepsilon_0,c_0)$$

$$= e^{TA}c_1 + e^{TA}\int_0^T e^{-sA}\{\varepsilon_0 F[s,x(s,\varepsilon_0,c_1),\varepsilon_0] + G(s)\}ds$$

$$- e^{TA}c_0 - e^{TA}\int_0^T e^{-sA}\{\varepsilon_0 F[s,x(s,\varepsilon_0,c_0),\varepsilon_0] + G(s)\}ds$$

$$= e^{TA}(c_1 - c_0) + e^{TA}\int_0^T e^{-sA}\{\varepsilon_0 B(s,\varepsilon_0)u(s,c_1 - c_0) \tag{29}$$

$$+ \varepsilon_0 H[u(s,c_1 - c_0)s,\varepsilon_0]\}ds$$

Also by equation (22), we have:

$$\tilde{F}(c_1,\varepsilon_0) - \tilde{F}(c_0,\varepsilon_0) = (e^{TA} - I)(c_1 - c_0)$$

$$+ e^{TA}\int_0^T e^{-sA}\{\varepsilon_0 B(s,\varepsilon_0)u(s,c_1 - c_0) \tag{30}$$

$$+ \varepsilon_0 H[u(s,c_1 - c_0),s,\varepsilon_0]\}ds$$

Equations (29) and (30) show that the differential at 0 of the mapping $\tilde{M}$, which we know already is $U(T)$, equals the differential at c_0 of $\tilde{F}(c, \varepsilon_0)$ plus the identity map. That is, if $D_0 \tilde{F}$ denotes the differential of $\tilde{F}(\cdot, \varepsilon_0)$ at c_0, then

$$U(T) = D_0 \tilde{M} = D_{c_0} \tilde{F} + I$$

Thus if $\lambda_1, \ldots, \lambda_n$ are the eigenvalues of $U(T)$, then $\lambda_1 - 1, \ldots, \lambda_n - 1$ are the eigenvalues of $D_{c_0} \tilde{F}$.

Our next object is to use the value of $\deg[M_0, \mathcal{B}, 0]$ to obtain information about the eigenvalues of $D_{c_0} \tilde{F}$. In order to do this, we impose further assumptions on the matrix A in equation (8).

Assumption 1: The matrix A is such that $e^{TA} - I$ has one of the following two forms:

$$\begin{bmatrix} 0 & \\ & C_1 \end{bmatrix}$$

or

$$\begin{bmatrix} 0 & 0 & \\ 0 & 0 & \\ & & C_2 \end{bmatrix}$$

where C_1 is an $(n-1) \times (n-1)$ nonsingular matrix, all of whose eigenvalues have negative real parts and C_2 is an $(n-2) \times (n-2)$ nonsingular matrix, all of whose eigenvalues have negative real parts.

(That is, we assume that

$$0 < n - r \le 2$$

and if $\lambda = 0$ is an eigenvalue of A, it has algebraic multiplicity 1 or 2 and if the multiplicity is 2, there are two linearly independent eigenvectors. If $\lambda = (2n\pi/T)i$ is an eigenvalue $(n = 1, 2, \ldots)$, its multiplicity is 1. In addition, we require that C_1, C_2 satisfy the conditions stated above.)

Suppose that

$$e^{TA} - I = \begin{bmatrix} 0 & \\ & 0 \\ & & C_2 \end{bmatrix}$$

Then a basis for E_{n-r} is the set of vectors

$$\begin{bmatrix} 1 \\ 0 \\ \vdots \\ 0 \end{bmatrix}, \quad \begin{bmatrix} 0 \\ 1 \\ 0 \\ \vdots \\ 0 \end{bmatrix}$$

and

$$P_{n-r} : (c_1, c_2, \ldots, c_n) \rightarrow (c_1, c_2)$$
$$P_r : (c_1, c_2, \ldots, c_n) \rightarrow (c_3, \ldots, c_n)$$

The matrix H is $\begin{bmatrix} 1 & & \\ & 1 & \\ & & [C_2^{-1}] \end{bmatrix}$.

Applying H to (22), we obtain:

$$H \tilde{F}(c_0, \varepsilon_0) = P_r c_0 + H e^{TA} \int_0^T e^{-sA} \{ \varepsilon_0 F[s, x(s, \varepsilon_0, c_0), \varepsilon_0]$$

$$+ G(s) \} ds = 0$$

This equation is equivalent to the following pair of equations:

$$P_{n-r} H e^{TA} \int_0^T e^{-sA} \{ \varepsilon_0 F[\quad] + G(s) \} ds = 0$$

$$P_r c_0 + P_r H e^{TA} \int_0^T e^{-sA} \{ \varepsilon_0 F[\quad] + G(s) \} ds = 0$$

or, using (12) and (15),

$$P_{n-r}H \int_0^T e^{-sA}\{\varepsilon_0 F[\quad]\}ds = 0$$

$$P_r c_0 + P_r H e^{TA} \int_0^T e^{-sA}\{\varepsilon_0 F[\quad] + G(s)\}ds = 0$$

and $\tilde{F}(c_0, \varepsilon_0)$ is described by

$$P_{n-r}H \int_0^T e^{-sA}\{\varepsilon_0 F[\quad]\}ds$$

$$C_2[P_r c_0] + C_2 P_r H e^{TA} \int_0^T e^{-sA}\{\varepsilon_0 F[\quad] + G(s)\}ds$$

where $C_2 P_r H = P_r$. Thus, if $|\varepsilon_0|$ is sufficiently small, the eigenvalues of $D_{c_0}\tilde{F}$ have real parts with the same sign as the eigenvalues of the matrix

$$\begin{bmatrix} \varepsilon_0 D_{c_0} M & \\ & C_2 \end{bmatrix}$$

where M denotes M_{ε_0}, the mapping defined after equation (20). Hence we are reduced to considering the eigenvalues of $D_{c_0}M$. At this point, we use information about the value of $\deg[M_0, \mathcal{B}, 0]$. Suppose that

$$\deg[M_0, \mathcal{B}, 0] > 0$$

If $F(t, x, \varepsilon)$ is changed by adding a "small" function $h(t)$ (i.e. if Theorem 7.4 is applied), then the Jacobian of M_0 is nonzero at each solution c of the equation

$$M_0(c) = 0$$

For convenience in this discussion we use the following notation: if f is a differentiable mapping from R^n into R^n and x is a point in the domain of f, let $J_x f$ denote the Jacobian of f at the point x, i.e. $J_x f$ denotes the determinant of $D_x f$, the differential of f at x.

Since $\deg[M_0, \mathcal{B}, 0] > 0$, then there is at least one point c_0 such that

$$J_{c_0} M_0 > 0$$

Suppose the eigenvalues of $D_{c_0} M_0$ are real. Since $J_{c_0} M_0$ is the product of the eigenvalues, then either both eigenvalues are positive or both are negative. If both eigenvalues are positive, and if ε_0 is positive, then both eigenvalues of

$$\varepsilon_0 D_{c_0} M$$

are positive. Hence, $D_{c_0} \tilde{F}$ has two eigenvalues with positive real parts and hence $U(T)$ has two real eigenvalues with real parts greater than one. Hence $x(t, \varepsilon_0, c_0)$ is unstable.

A similar argument shows that if both the eigenvalues of $D_{c_0} M_0$ are negative, then if ε_0 is positive, and sufficiently small, solution $x(t, \varepsilon_0, c_0)$ is asymptotically stable.

If

$$J_{c_0} M < 0$$

then both eigenvalues of $D_{c_0} M$ are real and one eigenvalue is positive and one is negative. The same sort of arguments as before show that $x(t, \varepsilon_0, c_0)$ is unstable.

If $J_{c_0} M_0 > 0$ and the eigenvalues of $D_{c_0} M_0$ are complex conjugates, i.e. the eigenvalues are $\alpha + i\beta$ and $\alpha = i\beta$, suppose first that $\alpha > 0$. Then if $\varepsilon_0 > 0$, both eigenvalues of $\varepsilon_0 D_{c_0} M$ have positive real parts and $D_{c_0} \tilde{F}$ has two eigenvalues with positive real parts. Hence, $U(T)$ has two eigenvalues whose real parts are greater than one. Hence, $x(t, \varepsilon_0, c_0)$ is unstable. If $\alpha < 0$, similar arguments show that if $\varepsilon_0 > 0$, then $x(t, \varepsilon_0, c_0)$ is asymptotically stable.

Similar arguments may be used if $\varepsilon_0 < 0$. We shall not consider the case in which the eigenvalues of $D_{c_0} M$ are $i\beta$ and $-i\beta$, i.e. we require that the trace of $D_{c_0} M$ be nonzero. Later, in looking at examples, we shall see how to obtain a computable sufficient condition that the trace of $D_{c_0} M$ be nonzero.

We may summarize our results in the following theorem.

Theorem 7.7 *If the matrix A satisfies Assumption 1 and $e^{TA} - I = \begin{bmatrix} 0 & 0 \\ & C_2 \end{bmatrix}$, if $\deg[M_0, \mathcal{B}, 0] \neq 0$, if the function $F(t, x, \varepsilon)$ in equation*

(8) *is replaced by* $F(t, x, \varepsilon) + h(t)$ *as described in the statement of Theorem 7.4 and if* $\operatorname{tr} D_{c_0} M \neq 0$ *at each point* c_0 *such that* $M(c_0) = 0$, *then the following conclusions hold:*

(1) *If* $\deg[M_0, \mathcal{B}, 0] < 0$, *then if* $\varepsilon > 0$ *and* ε *is sufficiently small, equation* (8) *has at least* $|\deg[M_0, \mathcal{B}, 0]|$ *distinct unstable periodic solutions.*

(2) *If* $\deg[M_0, \mathcal{B}, 0] > 0$, *then if* $|\varepsilon|$ *is sufficiently small, equation* (8) *has at least* $|\deg[M_0, \mathcal{B}, 0]|$ *distinct periodic solutions* $x(t, \varepsilon)$ *and each of these solutions* $x(t, \varepsilon)$ *has the following property: if* $\varepsilon > 0$ *and sufficiently small and both eigenvalues of* $D_{c_0} M_0$ *are positive [negative] solution* $x(t, \varepsilon)$ *is unstable [asymptotically stable]. If* $\varepsilon > 0$ *and sufficiently small and and the real part of each eigenvalue of* $D_{c_0} M_0$ *is positive [negative] solution* $x(t, \varepsilon)$ *is unstable [asymptotically stable].*

We leave the case

$$e^{TA} - I = \begin{bmatrix} 0 & \\ & C_1 \end{bmatrix}$$

as an exercise (Exercise 5).

Generalization of Malkin's Theorem

Now we prove a generalization of a theorem of Malkin [1959] concerning a particular class of resonance problems. The proof given here is much shorter and somewhat more transparent than that of Malkin. The purpose of the theorem is to obtain a branching result, which is a generalization of the branching result (Theorem 7.4) for the equation

$$x' = Ax + \varepsilon F(t, x, \varepsilon) + G(t) \tag{8}$$

That result said roughly that when ε is made nonzero, then periodic solutions branch from a finite set of the periodic solutions of (8).

In the generalization of Malkin's result to be described, the linear equation

$$x' = Ax$$

is replaced by a nonlinear equation

$$x' = G(t, x)$$

and we assume that this nonlinear equation has a continuous family of solutions. Then it is shown that when ε is made nonzero, periodic solutions branch from a finite number of the periodic solutions in the continuous family.

We assume that the equation

$$x' = G(x, t, \varepsilon) \tag{31}$$

satisfies the following hypothesis:

(i) $G: R^n \times R \times I \to R^n$ has continuous second derivatives in all variables at each point in $R^n \times R \times I$;

(ii) G has period T in t;

(iii) the equation

$$x' = G(x, t, 0)$$

has a k-parameter family of solutions

$$x(t, h_1, \ldots, h_k) = (x_1(t, h_1, \ldots, h_k)), \ldots, (x_n(t, h_1, \ldots, h_k))$$

all of period T, where $k \le n$ and $x(t, h_1, \ldots, h_k)$ has continuous first derivatives in $h_1, \ldots, h_k$;

(iv) the matrix

$$\begin{bmatrix} \frac{\partial x_1}{\partial h_1} & \cdots & \frac{\partial x_1}{\partial h_k} \\ \cdot & & \cdot \\ \frac{\partial x_n}{\partial h_1} & \cdots & \frac{\partial x_n}{\partial h_k} \end{bmatrix}$$

in which each entry is evaluated at $t = 0$, has rank k. For definiteness, assume that the matrix $[\partial x_i / \partial h_j]$, $i, j = 1, \ldots, k$, is nonsingular.

From assumption (iv), it follows that $\partial x/\partial h_1, \ldots, \partial x/\partial h_k$ are k linearly independent solutions of period T of the linear variational equation of (31) at $\varepsilon = 0$ and relative to $x(t, h_1, \ldots, h_k)$, i.e. the equation

$$y' = \{G_x[x(t, h_1, \ldots, h_k), t, 0]\}y \qquad (31, V)$$

We make one additional assumption:

(v) The dimension of the linear space of solutions of period T of equation (31.V) is equal to k.

We study the following problem.

Problem 4. If hypotheses (i), (ii), (iii), (iv) (v) hold and if $|\varepsilon|$ is sufficiently small, does there exist $(\bar{h}, \ldots, \bar{h}_k)$ such that equation (31) has a solution $y(t, \varepsilon)$ of period T with

$$\lim_{\varepsilon \to 0} y(t, \varepsilon) = x(t, \bar{h}_1, \ldots, \bar{h}_k)$$

for all $t \in [0, T]$.

We study Problem 4 by investigating solutions of equation (31) of the form

$$y(t, \varepsilon) = x(t, h_1, \ldots, h_k) + \varepsilon x(t, \varepsilon) \qquad (32)$$

Substituting from (32) into (31), we have

$$y' = G[y(t, \varepsilon), t, \varepsilon] \qquad (33)$$

or

$$x'(t, h_1, \ldots, h_k) + \varepsilon x'(t, \varepsilon) = G[x(t, h_1, \ldots, h_k) + \varepsilon x(t, \varepsilon), t, \varepsilon] \quad (34)$$

Subtracting the equation

$$x'(t, h_1, \ldots, h_k) = G[x(t, h_1, \ldots, h_k), t, 0] \qquad (35)$$

from (34), we have

$$
\begin{aligned}
\varepsilon x' &= G[x(t, h_1, \ldots, h_k) + \varepsilon x(t, \varepsilon), t, \varepsilon] - G[x(t, h_1, \ldots, h_k), t, 0] \\
&= \{G_x[x(t, h_1, \ldots, h_k), t, 0]\}\varepsilon x(t, \varepsilon) + \{G_\varepsilon[x(t, h_1, \ldots, h_k), t, 0]\}\varepsilon \\
&\quad + \varepsilon^2 \tilde{G}[x(t, h_1, \ldots, h_k), x(t, \varepsilon), t, \varepsilon]
\end{aligned}
\tag{37}
$$

or

$$
x' = \{B_x\}x + G_\varepsilon + \varepsilon\tilde{G}
\tag{38}
$$

where G_x, G_ε, $\tilde{G}$ have similar meanings to the corresponding terms g_u, g_ε and $\mathcal{G}$ in equation (7).

Let $\Phi(t)$ be a fundamental matrix for equation (31,V). By assumption (v), $\Phi(t)$ may be chosen so that $\Phi(0) = I$, the identity, and

$$
\Phi(T) = \begin{bmatrix} \begin{bmatrix} 1 & & & \\ & \cdot & & \\ & & \cdot & \\ & & & 1 \end{bmatrix} & \\ & \begin{bmatrix} B \end{bmatrix} \end{bmatrix}
$$

where the identity matrix is a $k \times k$ matrix and B is a nonsingular matrix which does not have an eigenvalue equal to one. (Hence $(B - I)^{-1}$ exists.) The condition that a solution

$$
y(t, w) = x(t, h_1, \ldots, h_k) + \varepsilon x(t, \varepsilon, c)
$$

of (38), where $y(0, w) = w$ and $\varepsilon x(0, \varepsilon, c) = \varepsilon c = w - x(0, h_1, \ldots, h_k)$, have period T is by the same considerations as in the derivation of equation (11),

$$
\begin{aligned}
[\Phi(T) - \Phi(0)]c + \int_0^T \Phi(T - s)\{G_\varepsilon[x(s, h_1, \ldots, h_k), s, 0] \\
+ \varepsilon\tilde{G}[x(s, h_1, \ldots, h_k), \varepsilon x(s, \varepsilon, c), s, \varepsilon]\}ds = 0
\end{aligned}
\tag{39}
$$

Let $\mathcal{P}_k, \mathcal{P}_{n-k}$ denote the projections:

$$\mathcal{P}_k: (x_1, \ldots, x_k, x_{k+1}, \ldots, x_n) \to (x_1, \ldots, x_k, 0, \ldots, 0)$$
$$\mathcal{P}_{n-k}: (x_1, \ldots, x_k, x_{k+1}, \ldots, x_n) \to (0, \ldots, 0, x_{k+1}, \ldots, x_n)$$

Then because of the conditions on $\Phi(T)$ and $\Phi(0)$, equation (39) can be represented by the pair of equations:

$$\mathcal{P}_k \int_0^T \Phi(T-s)\{G_\varepsilon + \varepsilon\tilde{G}\}ds = 0 \tag{40}$$

$$(B-I)(c_{k+1}, \ldots, c_n) + \mathcal{P}_{n-k} \int_0^T \Phi(T-s)\{G_\varepsilon + \varepsilon\tilde{G}\}ds = 0 \tag{41}$$

To find periodic solutions we must solve (40) and (41) for $c_1, \ldots, c_n$ in terms of ε. Setting $\varepsilon = 0$ and using the fact that there exists $(B-I)^{-1}$, we obtain from (41)

$$(c_{k+1}, \ldots, c_n) = -(B-I)^{-1}\mathcal{P}_k \int_0^T \Phi(T-s)\{G_\varepsilon\}ds$$

For brevity, denote this last expression by L. Thus, $(c_{k+1}, \ldots, c_n) = L$ and $\varepsilon = 0$ is an initial solution of (41) and since there exists $(B-I)^{-1}$, the Implicit Function Theorem can be applied to solve (41) uniquely for $(c_{k+1}, \ldots, c_n)$ in a neighborhood of L in terms of ε in a neighborhood of 0. The solution for $(c_{k+1}, \ldots, c_n)$ also depends differentiably on $h_1, \ldots, h_k$ and on $c_1, \ldots, c_k$. That is, we have

$$c_j = c_j(c_1, \ldots, c_k, h_1, \ldots, h_k, \varepsilon), \quad j = k+1, \ldots, n$$

Thus the problem of solving (40) and (41) has been reduced to the problem of solving the equation

$$\mathcal{P}_k \int_0^T \Phi(T-s)\{G_\varepsilon[x(s, h_1, \ldots, h_k), s, 0]$$
$$+ \varepsilon\tilde{G}[x(s, h_1, \ldots, h_k), x(s, \varepsilon, c_1, \ldots, c_k, c_{k+1}[c_1,$$
$$\ldots, c_k, h_1, \ldots, h_k, \varepsilon],$$
$$\ldots, c_n[c_1, \ldots, c_k, h_1, \ldots, h_k, \varepsilon]), s, \varepsilon]\}ds = 0 \tag{42}$$

Let

$$\tilde{B} = \mathcal{P}_k \int_0^T \Phi(T - s)\{G_\varepsilon[x(s, h_1, \ldots, h_k), s, 0]\}ds$$

Then (42) may be written:

$$\tilde{B} + \varepsilon\tilde{H}(c_1, \ldots, c_k, h_1, \ldots, h_k, \varepsilon) = 0$$

where $\tilde{H}$ has continuous first derivatives in all variables.

Now we want to point out that the last term on the left in equation (42) can be regarded as a function of $w_1, \ldots, w_n$, the components of w, the initial value of

$$y(t, w) = x(t, h_1, \ldots, h_k) + \varepsilon x(t, \varepsilon, c)$$

We write

$$\varepsilon x(t, \varepsilon, c) = y(t, w) - x(t, h_1, \ldots, h_k)$$

in the aforementioned term and use the fact that the solution $y(t, w)$ is a differentiable function of its initial value w. Now let

$$\tilde{P}(h_1, \ldots, h_k, \varepsilon) = \tilde{B} + \varepsilon\tilde{H}(0, \ldots, 0, h_1, \ldots, h_k, \varepsilon)$$

Then we have the matrix equation:

$$\left[\frac{\partial\tilde{P}_i}{\partial h_j}\right] = \left[\frac{\partial(\tilde{B} + \varepsilon\tilde{H})_i}{\partial w_\ell}\right]\left[\frac{\partial w_\ell}{\partial h_j}\right] \tag{43}$$

where $1 \leq \ell \leq k$. But

$$w = y(0, w) = x(0, h_1, \ldots, h_k) + \varepsilon x(0, \varepsilon, c)$$
$$= x(0, h_1, \ldots, h_k) + \varepsilon c$$

Hence

$$\left[\frac{\partial w_\ell}{\partial h_j}\right] = \left[\frac{\partial x_\ell(0, h_1, \ldots, h_k)}{\partial h_j}\right]$$

and by assumption (iv), this last matrix is nonsingular. Hence, by (43), the matrix

$$\left[\frac{\partial \tilde{P}_i}{\partial h_j}\right]$$

is nonsingular if and only if the matrix

$$\left[\frac{\partial(\tilde{B} + \varepsilon\tilde{H})_i}{\partial w_\ell}\right]$$

is nonsingular. But for $i = 1, \ldots, n$, we have by the chain rule

$$\frac{\partial}{\partial w_i}(\tilde{B} + \varepsilon\tilde{H}) = \frac{\partial}{\partial c_i}(\tilde{B} + \varepsilon\tilde{H})\frac{\partial c_i}{\partial w_i} \tag{44}$$

Since $w_i = x_i(0, h_1, \ldots, h_k) + \varepsilon c_i$, then $\partial c_i/\partial w_i = 1/\varepsilon$ and equation (44) becomes

$$\frac{\partial}{\partial w_i}(\tilde{B} + \varepsilon\tilde{H}) = \frac{1}{\varepsilon}\frac{\partial}{\partial c_i}(\tilde{B} + \varepsilon\tilde{H}) \tag{45}$$

But $\tilde{B}$ is independent of $c_1, \ldots, c_k$. Hence, $\frac{\partial \tilde{B}}{\partial c_i} = 0$ and equation (45) becomes

$$\frac{\partial}{\partial w_i}(\tilde{B} + \varepsilon\tilde{H}) = \frac{\partial}{\partial c_i}\tilde{H}$$

Hence matrix

$$\left[\frac{\partial \tilde{P}_i}{\partial h_j}\right]$$

is nonsingular if and only if the matrix

$$\left[\frac{\partial \tilde{H}_i}{\partial c_j}\right]$$

is nonsingular.

Summarizing, we have the following theorems.

Theorem 7.8 *Suppose that conditions* (i), (ii), (iii), (iv) *and* (v) *are satisfied; if* $(\bar{h}_1, \ldots, \bar{h}_k)$ *is such that*

$$\tilde{P}(\bar{h}_1, \ldots, \bar{h}_k, 0) = 0$$

and $[\partial \tilde{H}_i / \partial c_j](\bar{h}_1, \ldots, \bar{h}_k, 0)$ *is nonsingular, then if* $|\varepsilon|$ *is sufficiently small, equation* (31) *has a solution* $\bar{x}(t, \varepsilon)$ *of period* T *and*

$$\lim_{\varepsilon \to 0} \bar{x}(t, \varepsilon) = x(t, \bar{h}_1, \ldots, \bar{h}_k)$$

If $[\partial \tilde{H}_i / \partial c_j](\bar{h}_1, \ldots, \bar{h}_k, 0)$ is singular, then one may proceed to a study of the topological degree of the mapping described by the left side of (42) with $(h_1, \ldots, h_k) = (\bar{h}_1, \ldots, \bar{h}_k)$.

Examples of Branching of Periodic Solutions

Now we describe some examples of system (8), derive the corresponding mapping M_0 and compute its degree. We start with a class of 2-dimensional systems which often arise in practice. Let us assume that

$$A = \begin{bmatrix} 0 & \beta \\ -\beta & 0 \end{bmatrix}$$

where, for simplicity of notation, we take $\beta = 2\pi$. Then $E_{n-r} = R^2$, $P_{n-r} = I$, the identity 2×2 matrix, and $P_r = 0$. Also

$$e^{tA} = \begin{bmatrix} \cos \beta t & \sin \beta t \\ -\sin \beta t & \cos \beta t \end{bmatrix}$$

which has period $T = 1$ and

$$e^{TA} - I = \begin{bmatrix} 0 & 0 \\ 0 & 0 \end{bmatrix}$$

Since $P_r = 0$, the matrix H can be taken to be I, the identity. This is, of course, an example of a resonance case. If

$$G(t) = \begin{bmatrix} G_1(t) \\ G_2(t) \end{bmatrix}$$

then the necessary condition stated in Theorem 7.2, i.e. equation (15), becomes:

$$\int_0^T \begin{bmatrix} \cos 2\pi s & -\sin 2\pi s \\ \sin 2\pi s & \cos 2\pi s \end{bmatrix} \begin{bmatrix} G_1(s) \\ G_2(s) \end{bmatrix} ds = 0 \qquad (45)$$

That is, the vector $G(t)$ is orthogonal, in a conventional sense, to the vectors

$$\begin{bmatrix} \cos 2\pi t \\ -\sin 2\pi t \end{bmatrix} \quad \text{and} \quad \begin{bmatrix} \sin 2\pi t \\ \cos 2\pi t \end{bmatrix}$$

In other words, vector $G(t)$ is orthogonal to the linear space of solutions of the equation

$$x' = Ax$$

Notice that condition (45) may not be satisfied if $G(t)$ is a function with period 1, e.g., if

$$G(t) = \begin{bmatrix} \cos 2\pi t \\ -\sin 2\pi t \end{bmatrix}$$

On the other hand, if $G(t)$ has period $1/2$, e.g., if

$$G(t) = \begin{bmatrix} \cos 4\pi t \\ -\sin 4\pi t \end{bmatrix}$$

the condition (45) is satisfied. The forcing term $G(t)$, which describes a periodic force imposed from the outside on the physical system described by

$$x' = Ax + \varepsilon F(t, x, \varepsilon)$$

often produces an oscillation of the system (i.e. the equation

$$x' = Ax + \varepsilon F(t, x, \varepsilon) + G(t) \qquad (8)$$

has a periodic solution) if the resonance case is being considered and if the frequency of the forcing term is an integral multiple of the frequency of a periodic solutions of $x' = Ax$ (i.e. the period of some periodic solution of $x' = Ax$ is an integral multiple of the period of $G(t)$). The resulting oscillation of the system (periodic solution of (8)) is sometimes called a subharmonic oscillation.

The mapping M_0 in this case is (see equation (21)):

$$M_0: (c_1, c_2) \rightarrow \int_0^1 e^{-sA} F[s, e^{sA}c, 0]\, ds$$

where $c = \begin{bmatrix} c_1 \\ c_2 \end{bmatrix}$. If the components of F are F_1 and F_2, then we have:

$$\int_0^1 e^{-sA} F[s, e^{sA}c, 0]\, ds = \int_0^1 \begin{bmatrix} \cos \beta s & -\sin \beta \\ \sin \beta s & \cos \beta s \end{bmatrix} \begin{bmatrix} F_1(\ ) \\ F_2(\ ) \end{bmatrix} ds$$

where

$$F_i(\ ) = F_i(s, c_1 \cos \beta s + c_2 \sin \beta s, -c_1 \sin \beta s + c_2 \cos \beta s, 0)$$

Thus

$$M_0: (c_1, c_2) \rightarrow \begin{bmatrix} \int_0^1 [(\cos \beta s)F_1(\ ) - (\sin \beta s)F_2(\ )]ds \\ \int_0^1 [(\sin \beta s)F_1(\ ) + (\cos \beta s)F_2(\ )]ds \end{bmatrix}$$

Now let us assume that $F_i(s, \xi_1, \xi_2, 0)$ is a polynomial in ξ_1 and ξ_2 and is independent of s. From the fact that

$$\int_0^1 (\sin^m x)(\cos^n x)\, dx \neq 0$$

if and only if m and n are both even integers, it follows that

$$\int_0^1 \begin{bmatrix} \cos \beta s & -\sin \beta s \\ \sin \beta s & \cos \beta s \end{bmatrix} \begin{bmatrix} F_1(\ ) \\ F_2(\ ) \end{bmatrix} ds = \begin{bmatrix} P_1(c_1, c_2) \\ P_2(c_1, c_2) \end{bmatrix}$$

where $P_1(c_1, c_2)$ and $P_2(c_1, c_2)$ are polynomials in c_1 and c_2 and each term in P_1 and P_2 is of the form

$$K c_1^{q_1} c_2^{q_2}$$

where K is a constant coefficient and $q_1 + q_2$ is an odd number. Suppose that

$$P_1(c_1, c_2) = k_1 c_1 + k_2 c_2 + H_1(c_1, c_2)$$
$$P_2(c_1, c_2) = k_3 c_1 + k_4 c_2 + H_2(c_1, c_2)$$

where

$$\det \begin{bmatrix} k_1 & k_2 \\ k_3 & k_4 \end{bmatrix} \neq 0$$

and for $i = 1, 2$, the polynomial $H_i(c_1, c_2)$ consists of terms of the form $K c_1^{q_1} c_2^{q_2}$ where $q_1 + q_2 \geq 2$. Then by Example 8 in the Appendix if the radius of $\mathcal{B}$ is sufficiently small,

$$\deg[M_0, \mathcal{B}, 0] = \text{sgn} \det \begin{bmatrix} k_1 & k_2 \\ k_3 & k_4 \end{bmatrix}$$

Suppose that

$$P_1(c_1, c_2) = p_1(c_1, c_2) + q_1(c_1, c_2)$$
$$P_2(c_1, c_2) = p_2(c_1, c_2) + q_2(c_1, c_2)$$

where $p_1(c_1, c_2)$ is a polynomial homogeneous of degree s_1 in c_1 and c_2, $p_2(c_1, c_2)$ is a polynomial homogeneous of degree s_2 in c_1, c_2, and for $i = 1, 2$, $q_i(c_1, c_2)$ consists of terms of the form $k c_1^{\ell_1^{(i)}} c_2^{\ell_2^{(i)}}$ where $\ell_1^{(i)} + \ell_2^{(i)} < \min(s_1, s_2)$. Let $\bar{M}$ be the mapping defined by:

$$\bar{M} : (c_1, c_2) \rightarrow (p_1(c_1, c_2), p_2(c_1, c_2))$$

Assume that p_1 and p_2 have no common real linear factors. Then $\deg[\bar{M}, \mathcal{B}, 0]$ is defined for $\mathcal{B}$ of arbitrary radius and by Example 13 in the Appendix if the radius of $\mathcal{B}$ is sufficiently large, then

$$\deg[M_0, \mathcal{B}, 0] = \deg[\bar{M}, \mathcal{B}, 0]$$

Moreover, $\deg[\bar{M}, \mathcal{B}, 0]$ can be computed explicitly as shown in Example 15 of the Appendix.

After $\deg[M_0, \mathcal{B}, 0]$ has been computed, then Theorem 7.4 shows that if a "small" function $h(t)$ is added to $F(t, x, \varepsilon)$, then $|\deg[M_0, \mathcal{B}, 0]|$ is a lower bound for the number of periodic solutions.

Next we want to look at the stability of the periodic solutions thus obtained, i.e. we want to apply Theorem 7.7. In order to apply Theorem 7.7 to equation (8), it is sufficient to impose a hypothesis which implies that $\operatorname{tr} D_c M \neq 0$ for all $c \in \mathcal{B}$. Thus if for all $c \in \mathcal{B}$,

$$\frac{\partial p_1}{\partial c_1} + \frac{\partial p_2}{\partial c_2} \neq 0$$

then Theorem 7.7 is applicable to (8).

If we consider an example of equation (8) in which the dimensions of the system are greater than two, the procedure for computing the mapping M_0 is very much the same except that it is more complicated. In the proof of Theorem 7.4, we computed a typical such mapping M_0.

B. Bifurcation of Periodic Solutions

The General Problem and a Partial Solution to It

We return to the study of the general equation

$$u' = f(t, u, \varepsilon) \tag{1}$$

except that now we consider the autonomous case. We assume that f is independent of t so that we deal with the equation

$$u' = f(u, \varepsilon) \tag{1'}$$

where the n-vector function f has continuous third derivatives at each point

$$(u, \varepsilon) \in R^n \times I$$

where I is an open interval on the real line with midpoint zero and for $\varepsilon = 0$, equation (1') has a solution $u(t)$ of period T.

We study the following problem:

Problem 1. Does there exist a positive-valued twice differentiable function $T(\varepsilon)$ with domain an open interval J with $0 \in J \subset I$ such that $T(0) = T$ and such that if $|\varepsilon|$ is sufficiently small, equation (1) has a solution $u(t,\varepsilon)$ of period $T(\varepsilon)$, with

$$\lim_{\varepsilon \to 0} u(t,\varepsilon) = u(t)$$

for each real t.

Notice that even the phrasing of the basic problem is different in the nonautonomous and autonomous cases. In the nonautonomous case we are given the function $T(\varepsilon)$ which is the period in t of $f(t,x,\varepsilon)$. In the autonomous case, we must search for the function $T(\varepsilon)$. It is rather natural to ask why we should not simply choose arbitrarily a function $T(\varepsilon)$. For example, why not choose $T(\varepsilon) = T$ for all $\varepsilon \in I$? There is an important physical reason for not specifying the function $T(\varepsilon)$. We have assumed that the equation

$$u' = f(u,0)$$

has a solution of period T. If the equation describes a physical system, then the physical system has a "natural" oscillatory period equal to T. But when the system is perturbed by an influence described by

$$f(u,\varepsilon) - f(u,0)$$

i.e. when the system is perturbed so that it is described by the equation

$$u' = f(u,0) + [f(u,\varepsilon) - f(u,0)] = f(u,\varepsilon)$$

there is no reason why, if the perturbed system has any natural oscillatory period at all, the natural period should continue to be equal to T or, more generally, there is no reason why we should be able to determine in advanced the natural period of oscillation of the perturbed system. A simple example of this is a pendulum with small oscillations, i.e. a pendulum in which the amplitudes of the oscillations are so small that the motion of the pendulum can be described by a linear ordinary differential equation. By elementary arguments it can be shown that the period of the pendulum (its "natural" period) depends on the length of the pendulum. Thus if the length of the pendulum is changed, the "natural" period changes.

In the treatment of equation (1'), we proceed as follows. We seek a twice differentiable function $T(\varepsilon)$, and we first write

$$T(\varepsilon) = T(0) + \varepsilon m(\varepsilon)T(0)$$

where $m(\varepsilon)$ is a differentiable function of ε. Let

$$s = (1 + \varepsilon m(\varepsilon))^{-1}t$$

and rewrite (1') as

$$\frac{du}{ds} = \frac{du}{dt} \cdot \frac{dt}{ds} = [1 + \varepsilon m(\varepsilon)][f(u,\varepsilon)] \tag{2}$$

We seek a solution of (2) of the form:

$$u(s) + \varepsilon x(s,\varepsilon) \tag{3}$$

where $u(s)$ is the given solution of period $T = T(0)$ of the equation

$$u' = f(u,0) \tag{4}$$

Substituting from (3) into (2), we obtain

$$
\begin{aligned}
u' + \varepsilon x' &= (1 + \varepsilon m(\varepsilon))f[u(s) + \varepsilon x(s,\varepsilon),\varepsilon] \\
&= (1 + \varepsilon m(\varepsilon))\{f[u(s),0] + \{f_u[u(s),0]\}\varepsilon x(s,\varepsilon) \\
&\quad + \varepsilon f_\varepsilon[u(s),0] + \varepsilon^2 F[u(s),x(s,\varepsilon),\varepsilon]\}
\end{aligned} \tag{5}
$$

and subtracting (4) from (5) yields:

$$
\begin{aligned}
\varepsilon x' &= (1 + \varepsilon m(\varepsilon))\{f_u[u(s),0]\varepsilon x(s,\varepsilon) + \varepsilon f_\varepsilon[u(s),0] \\
&\quad + \varepsilon^2 F[u(s),x(s,\varepsilon),\varepsilon]\} + \varepsilon m(\varepsilon)f[u(s),0]
\end{aligned} \tag{6}
$$

Dividing (6) by ε and rearranging terms, we obtain

$$
\begin{aligned}
x' &= f_u[u(s),0]x + \varepsilon m(\varepsilon)f_u[u(s),0]x + [\varepsilon + \varepsilon^2 m(\varepsilon)]F[u(s),x(s,\varepsilon),\varepsilon] \\
&\quad + [1 + \varepsilon m(\varepsilon)]f_\varepsilon[u(s),0] + m(\varepsilon)f[u(s),0]
\end{aligned} \tag{7}
$$

Thus, Problem 1 becomes:

Problem 2. To determine if there exists a differentiable function $m(\varepsilon)$ such that equation (7) has a solution $x(s, \varepsilon)$ of period $T(0)$.

This last, however, is not yet the desired formulation of the problem. The problem is complicated by the fact that we are concerned with solutions of the autonomous equation (1′). Since the desired solution

$$u(s) + \varepsilon x(s, \varepsilon)$$

is to be the solution of an autonomous equation, then we can reparameterized the solution (see Lemma 3.1), in any convenient way. Suppose that $u_1(s)$ is the first component of $u(s)$ and that $u_1'(0) \neq 0$ (at least one component of $u'(0)$ is nonzero because otherwise $u(0)$ would be an equilibrium point). Let

$$u_1(0) = \bar{u}_1$$

By choosing the right parameterization, we may require that the solution

$$u(s) + \varepsilon x(s, \varepsilon)$$

have the property that

$$u_1(0) + \varepsilon x_1(0, \varepsilon) = \bar{u}_1$$

and hence that

$$x_1(0, \varepsilon) = 0$$

In order to prove this can be done, let

$$g(t, \varepsilon) = u_1(t) + \varepsilon x_1(t, \varepsilon) - \bar{u}_1$$

Then

$$g(0,0) = 0$$

and

$$\frac{dg}{dt}(0,0) \neq 0$$

Hence by the Implicit Function Theorem, there exists a continuous function $\tau(\varepsilon)$ such that $\tau(0) = 0$ and

$$g(\tau(\varepsilon), \varepsilon) = 0$$

or

$$u_1(\tau(\varepsilon)) + x_1(\tau(\varepsilon), \varepsilon) = \bar{u}_1$$

The solution with the desired parameterization is thus

$$u_1(t + \tau(\varepsilon)) + \varepsilon x_1(t + \tau(\varepsilon), \varepsilon)$$

Thus, Problem 2 can be formulated more precisely as:

Problem 3. Can the function $m(\varepsilon)$ be chosen so that if $|\varepsilon|$ is sufficiently small, equation (7) has a solution $x(s, \varepsilon)$ such that

$$x_1(0, \varepsilon) = 0$$

and such that $x(s, \varepsilon)$ has period $T(0)$?

It will be more convenient in the discussion which follows to replace $m(\varepsilon)$ by m.

First, by Floquet theory, we change variables so that $f_u[u(s), 0]$ becomes a constant matrix A. Then equation (7) becomes (using the same letter x for the dependent variable and using t for the independent variable):

$$\begin{aligned}
x' = Ax &+ \varepsilon m A x + [\varepsilon + \varepsilon^2 m]\tilde{G}[u(t), x, \varepsilon, t] \\
&+ (1 + \varepsilon m)g[u(t), t] + m\bar{g}[u(t), t]
\end{aligned} \tag{8}$$

where $\tilde{G}$, g and $\bar{g}$ are continuously differentiable functions of the indicated variables and are periodic of period $T(0)$ as functions of t.

Now suppose that $u(t)$ is a nontrivial solution. Differentiating

$$u' = f(u, 0)$$

with respect to t, we have:

$$\frac{d^2 u}{dt^2} = \frac{d}{dt} u' = \frac{\partial f}{\partial u}[u(t), 0] \frac{du}{dt}$$

i.e. du/dt is a solution of period T of the linear variational equation

$$w' = \{f_u[u(s), 0]\} w$$

and hence the matrix $f_u[u(t), 0]$ has at least one characteristic multiplier equal to one, i.e. the matrix A has at least one eigenvalue equal to $2n\pi i/T$ $(n = 0, +1, \dots)$. Let us assume that the linear space of periodic solutions of the linear variational equation has dimension one, i.e. $f_u[u(t), 0]$ has the number one as a simple characteristic multiplier or matrix A has zero as a simple eigenvalue and A has no other eigenvalues of the form $(2n\pi/T)i$ where n is a nonnegative integer. This corresponds to the case in the branching problem where the matrix A has no eigenvalues of the form $\left(\frac{2n\pi}{T}\right)i$ with $n = 0, \pm 1, \pm 2, \dots$. In analogy with Theorem 7.1, we will prove:

Theorem 7.9 *If $f_u[(t), 0]$ has the number one as a simple characteristic multiplier, then if $|\varepsilon|$ is sufficiently small, there is a solution $u(t, \varepsilon)$ of $(1')$ such that $u(t, \varepsilon)$ has period $T(\varepsilon)$ and*

$$\lim_{\varepsilon \to 0} u(t, \varepsilon) = u(t)$$

for all real t.

Proof. By translation of axes, we may assume that the given periodic solution $u(t)$ is such that $u_1(0) = 0$ and by rotation of axes, we may assume

$$u_1'(0) \neq 0$$
$$u_2'(0) = \cdots = u_n'(0) = 0$$

Let $u(t, c, \varepsilon)$ denote the solution of $(1')$ such that

$$u(0, c, \varepsilon) = c$$

and, as pointed out earlier, we may choose

$$c = \begin{bmatrix} 0 \\ c_2 \\ \vdots \\ c_n \end{bmatrix}$$

since $u_1(0) = \bar{u}_1 = 0$ in this case. The periodicity condition is:

$$u(T(\varepsilon), c, \varepsilon) - u(0, c, \varepsilon) = 0$$

For the proof of this theorem, it is sufficient to write

$$T(\varepsilon) = T(0) + \tau(\varepsilon)$$

where $\tau(0) = 0$ and we write

$$c = u(0) + d = \begin{bmatrix} u_1(0) \\ u_2(0) \\ \vdots \\ u_n(0) \end{bmatrix} + \begin{bmatrix} 0 \\ d_2 \\ \vdots \\ d_n \end{bmatrix} = \begin{bmatrix} 0 \\ u_2(0) + d_2 \\ \vdots \\ u_n(0) + d_n \end{bmatrix}$$

Then the periodicity condition becomes:

$$u[T(0) + \tau(\varepsilon), c, \varepsilon] - u(0, c, \varepsilon) = 0$$

or

$$u[T(0) + \tau(\varepsilon), u(0) + d, \varepsilon] - (u(0) + d) = 0 \qquad \text{(P)}$$

We want to solve (P) for $\tau, d_2, \ldots, d_n$ as functions of ε by using the Implicit Function Theorem. First, equation (P) has the initial solution:

$$\varepsilon = 0, \ \tau = 0, \ d_2 = 0, \ldots, d_n = 0$$

and it is therefore sufficient to prove that the appropriate Jacobian is nonzero. The appropriate Jacobian is the determinant of the matrix:

$$\begin{bmatrix} u_1'(0) & \frac{\partial u_1}{\partial d_2}(T,0,0) & \cdots & \frac{\partial u_1}{\partial d_n} \\ 0 & \frac{\partial u_2}{\partial d_2} - 1 & \cdots & \frac{\partial u_2}{\partial d_n} \\ \vdots & & & \\ 0 & \frac{\partial u_n}{\partial d_2} & & \frac{\partial u_n}{\partial d_n} - 1 \end{bmatrix}$$

Since $u_1'(0) \neq 0$, it is sufficient to prove that

$$\det \begin{bmatrix} \frac{\partial u_2}{\partial d_2} - 1 & \cdots & \frac{\partial u_2}{\partial d_n} \\ \cdot & \cdot & \cdot \\ \frac{\partial u_n}{\partial d_2} & \cdots & \frac{\partial u_n}{\partial d_n} - 1 \end{bmatrix} \neq 0$$

If $\bar{u}(s,c,\varepsilon)$ denotes the solution of $(1')$ such that

$$\bar{u}(0,c,\varepsilon) = c$$

where

$$c = \begin{bmatrix} d_1 \\ d_2 \\ \vdots \\ d_n \end{bmatrix}$$

and d_1 is arbitrary, then a fundamental matrix of the linear variational system

$$w' = [f_u(u(t),0)]w$$

is:

$$U(t) = \begin{bmatrix} \frac{\partial \bar{u}_1}{\partial d_1} & \frac{\partial \bar{u}_1}{\partial d_2} & \cdots & \frac{\partial \bar{u}_1}{\partial d_n} \\ \frac{\partial \bar{u}_2}{\partial d_1} & \cdot & & \cdot \\ \vdots & & \cdot & \\ \frac{\partial \bar{u}_n}{\partial d_1} & \frac{\partial \bar{u}_n}{\partial d_2} & & \frac{\partial \bar{u}_n}{\partial d_n} \end{bmatrix}$$

where all the entries are evaluated at $(t, u(0), 0)$. Also $U(0) = I$, the identity, because by the Mean Value Theorem:

$$\bar{u}_1(t,c,0) = \bar{u}_1(0,u(0),0) + tm_1(t,c,0) = d_1 + tm_1(t,c,0)$$
$$\bar{u}_j(t,c,0) = d_j + tm_j(t,c,0) \qquad (j = 2,\ldots,n)$$

where m_k $(k = 1, \ldots, n)$ is a differentiable function in all variables. Hence

$$\frac{\partial \bar{u}_1}{\partial d_1}(0, u(0), 0) = 1$$

$$\frac{\partial \bar{u}_j}{\partial d_1}(0, u(0), 0) = 0 \qquad (j = 2, \ldots, n)$$

Similarly,

$$\frac{\partial \ddot{u}_j}{\partial d_k}(0, u(0), 0) = \delta_{jk} \qquad (j = 1, \ldots, n; \; k = 2, \ldots, n)$$

Since

$$\begin{bmatrix} \frac{\partial \bar{u}_1}{\partial d_1}(t, u(0), 0) \\ \vdots \\ \frac{\partial \bar{u}_n}{\partial d_1}(t, u(0),) \end{bmatrix}$$

is a solution of the linear variational equation and since

$$\begin{bmatrix} \frac{\partial \bar{u}_1}{\partial d_1}(0, u(0), 0) \\ \vdots \\ \frac{\partial \bar{u}_n}{\partial d_1}(0, u(0), 0) \end{bmatrix} = \begin{bmatrix} 1 \\ 0 \\ \vdots \\ 0 \end{bmatrix}$$

and since

$$\begin{bmatrix} \frac{\partial u_1}{\partial t}(t, u(0), 0) \\ \vdots \\ \frac{\partial u_n}{\partial t}(t, u(0), 0) \end{bmatrix}$$

is a solution of the linear variational equation with

$$\begin{bmatrix} \frac{\partial u_1}{\partial t}(0, u(0), 0) \\ \vdots \\ \frac{\partial u_n}{\partial t}(0, u(0), 0) \end{bmatrix} = \begin{bmatrix} u_1'(0) \\ 0 \\ \vdots \\ 0 \end{bmatrix}$$

it follows by uniqueness of solution that

$$\begin{bmatrix} \frac{\partial \bar{u}_1}{\partial d_1}(t, u(0), 0) \\ \vdots \\ \frac{\partial \bar{u}_n}{\partial d_1}(t, u(0), 0) \end{bmatrix} = \frac{1}{u_1'(0)} \begin{bmatrix} \frac{\partial u_1}{\partial t}(t, u(0), 0) \\ \vdots \\ \frac{\partial u_n}{\partial t}(t, u(0), 0) \end{bmatrix}$$

Hence, since $\frac{\partial u}{\partial t}(t, u(0), 0)$ has period T, it follows that

$$\frac{\partial \bar{u}_j}{\partial d_1}(0, u(0),) = \frac{\partial \bar{u}_j}{\partial d_1}(T, u(0), 0) \qquad (j = 1, \dots, n)$$

Hence

$$U(T) = \begin{bmatrix} 1 & \frac{\partial \bar{u}_1}{\partial d_2}(T, 0, 0) & \cdots & \frac{\partial \bar{u}_1}{\partial d_n} \\ 0 & \frac{\partial \bar{u}_2}{\partial d_2} & \cdot & \frac{\partial \bar{u}_2}{\partial d_n} \\ \vdots & \cdot & \cdot & \cdot \\ 0 & & & \frac{\partial \bar{u}_n}{\partial d_n} \end{bmatrix}$$

But by hypothesis the number 1 is a simple root of the equation

$$\det[U(T) - \lambda I] = 0$$

Hence

$$\det \begin{bmatrix} \frac{\partial \bar{u}_2}{\partial d_2}(T, 0, 0) - 1 & \cdots & \frac{\partial \bar{u}_2}{\partial d_2} \\ \cdot & \cdot & \cdot \\ \frac{\partial \bar{u}_n}{\partial d_1} & & \frac{\partial \bar{u}_n}{\partial d_n} - 1 \end{bmatrix} \neq 0$$

This completes the proof of Theorem 7.10 □

Next we consider the case in which the matrix A in equation (8) has at least two eigenvalues equal to $2n\pi i/T$ $(n = 0, \pm 1, \dots)$, i.e. the null space of $e^{AT} - I$ has dimension greater than or equal to two. Since

$$e^{t(1+\varepsilon m)A}$$

is a fundamental matrix of

$$x' = (1 + \varepsilon m)Ax$$

and this fundamental matrix is the identity at $t = 0$, then the condition that a solution $x(t, \varepsilon)$ of equation (8) have period $T = T(0)$ is

$$e^{T(1+\varepsilon m)A}c - c + e^{T(1+\varepsilon m)A}\int_0^T e^{-s(1+\varepsilon m)A}$$

$$\left\{[\varepsilon + \varepsilon^2 m]\tilde{G} + (1 + \varepsilon m)g + m\bar{g}\right\}ds = 0 \tag{9}$$

Now equation (9) may be rewritten as:

$$(e^{T(1+\varepsilon m)A} - e^{TA})c + (e^{TA} - I)c + e^{T(1+\varepsilon m)A}\int_0^T e^{-s(1+\varepsilon m)A}$$

$$\left\{[\varepsilon + \varepsilon^2 m]\tilde{G} + \tilde{H}\right\}ds = 0 \tag{10}$$

where $\tilde{H}(\varepsilon) = (1+\varepsilon m)g[u(s), s] + m\bar{g}[u(s), s]$. Applying the operator H (which is such that $H(e^{TA} - I)c = P_r c$) to equation (10), we obtain

$$H(e^{T(1+\varepsilon m)A} - e^{TA})c + P_r c + H e^{T(1+\varepsilon m)A}\int_0^T e^{-s(1+\varepsilon m)A}$$

$$\left\{[\varepsilon + \varepsilon^2 m]\tilde{G} + \tilde{H}\right\}ds = 0 \tag{11}$$

Applying P_r and P_{n-r} to (11), we have

$$P_r H(e^{T(1+\varepsilon m)A} - e^{TA})c + P_r c + P_r H e^{T(1+\varepsilon m)A}\int_0^T e^{-s(1+\varepsilon m)A}$$

$$\left\{[\varepsilon + \varepsilon^2 m]\tilde{G} + \tilde{H}\right\}ds = 0 \tag{12}$$

and

$$P_{n-r}H(e^{T(1+\varepsilon m)A} - e^{TA})c + P_{n-r}H e^{T(1+\varepsilon m)A}\int_0^T e^{-s(1+\varepsilon m)A}$$

$$\left\{[\varepsilon + \varepsilon^2 m]\tilde{G} + \tilde{H}\right\}ds = 0 \tag{13}$$

We assume that the following condition is satisfied:

Condition C. $\int_0^T e^{-sA}g[u(s),s]ds = 0$ and $\int_0^T e^{-sA}\{\bar{g}[u(s),s]\}ds = 0$.

(By following the reasoning used in the proof of Theorem 7.2, a weaker condition, which is a necessary condition to solve Problem 3, can be obtained. For simplicity in computation, we omit this.)

From (12), we see that if $\varepsilon = 0$, $P_r c = 0$. Dividing (13) by ε, we obtain

$$
P_{n-r}H \left[\frac{e^{T(1+\varepsilon m)A} - e^{TA}}{\varepsilon m} \frac{\varepsilon m}{\varepsilon} \right] c
$$

$$
+ P_{n-r}H e^{T(1+\varepsilon m)A} \int_0^T e^{-s(1+\varepsilon m)A} \left\{ [1 + \varepsilon m]\tilde{G} + \frac{\tilde{H}}{\varepsilon} \right\} ds = 0
$$

(14)

Using condition C and letting $\varepsilon \to 0$ in the left-hand side of (14), we obtain:

$$
\mathcal{M}(m,c) \stackrel{\text{def}}{=} P_{n-r}H e^{TA}TAmc + P_{n-r}H e^{TA} \int_0^T e^{-sA}\{\tilde{G}\}ds
$$

(15)

Let

$$
c = (0, c_2, \ldots, c_{n-r}, 0, \ldots, 0)
$$

Then $\mathcal{M}(m,c)$ defines a mapping

$$
\mathcal{M} : (m, c_2, \ldots, c_{n-r}) \to R^{n-r}
$$

By the Implicit Function Theorem, if $c_2, \ldots, c_{n-r}, m$ are arbitrary and $c_1 = 0$, equation (12) can be solved uniquely for $c_{n-r+1}, \ldots, c_n$ in terms of $c_2, \ldots, c_{n-r}, m, \varepsilon$ in neighborhoods of 0, i.e. we obtain

$$
c_j = \varepsilon C_j(c_2, \ldots, c_{n-r}, m, \varepsilon) \qquad (j = n - r + 1, \ldots, c_n) \qquad (16)
$$

where each function C_j has continuous first derivatives in all variables. Substituting from (16) into (14), we conclude that the resulting equation can be written as:

$$
\mathcal{M}(m, c_2, \ldots, c_{n-r}) + \varepsilon \mathcal{N}(\varepsilon, m, c_2, \ldots, c_{n-r}) = 0 \qquad (17)
$$

where $\mathcal{N}$ has continuous first derivatives in all variables. Thus in order to establish the existence of periodic solutions, it is sufficient to solve (17) for $m, c_2, \ldots, c_{n-r}$ as functions of ε. This is done by using the Implicit Function Theorem or studying the topological degree of the mapping

$$\mathcal{M} : (m, c_2, \ldots, c_{n-r}) \rightarrow R^{n-r}$$

The general theorems obtained in this way are long and awkward to state. (For application of the Implicit Function Theorem, see, e.g. Coddington and Levinson [1955].) Moreover the theorems are clumsy because in order to apply them to particular equations, it is usually necessary to go through the same kind of steps that are used to prove the general theorems. Hence we will omit the statements of theorems and, instead, illustrate the method with an example (Exercise 8).

The Hopf Bifurcation Theorem

In the results just described, a periodic solution is obtained for $|\varepsilon|$ small and $\varepsilon \neq 0$. In certain applications in biology and chemistry (see Kopell and Howard [1973]) it is important to be able to show that periodic solutions occur if $\varepsilon > 0$, or periodic solutions occur if $\varepsilon < 0$, Elegant and explicit results in this direction are given by the Hopf Bifurcation Theorem. First proved by E. Hopf [1942], this theorem has been extended and generalized by a number of writers. See Ruelle and Takens [1971] and Ize [1976]. For a collection of studies of the theorem, see Marsden and McCracken [1976]. We will describe here a simple version of the theorem which requires for its proof only the implicit function theorem (I am indebted to Professor Roger Nussbaum for this proof.)

We consider the n-dimensional equation

$$x' = Ax + \varepsilon L(\varepsilon)x + f(x, \varepsilon) \tag{18}$$

where

$$A = \begin{bmatrix} \begin{matrix} 0 & 1 \\ -1 & 0 \end{matrix} & \\ & D \end{bmatrix}$$

where D is an $(n-2) \times (n-2)$ matrix which has no eigenvalue of the form iq where q is an integer, $L(\varepsilon)$ is an $n \times n$ matrix, each entry of which is a differentiable function of ε for $\varepsilon \in J$, an open interval containing zero, and

$$L(0) = \begin{bmatrix} 1 & 0 & \\ 0 & 1 & \\ & & \bar{D} \end{bmatrix}, \quad L(\varepsilon) = \begin{bmatrix} h(\varepsilon) & 0 & \\ 0 & h(\varepsilon) & \\ & & \mathcal{D}(\varepsilon) \end{bmatrix}$$

where $\bar{D}$ is a real $(n-2) \times (n-2)$ matrix. Finally we assume that $f(x,\varepsilon)$ has continuous third derivatives at each point of $R^n \times J$ and that

$$|f(x,\varepsilon)| = o(|x|)$$

uniformly for $\varepsilon \in J$, i.e. for $\varepsilon \in J$ and $x \in R^n$

$$|f(x,\varepsilon)| < \eta(|x|)|x|$$

where $\eta(\tau)$ is a continuous positive-valued function with domain $\{\tau / \tau \geq 0\}$ and $\eta(0) = 0$.

Let $x(t,c,\varepsilon)$ denote the solution of (18) such that

$$x(0,c,\varepsilon) = c$$

and let λ be a positive number. The solution $x(t,c,\varepsilon)$ has period $2\pi\lambda$ if and only if:

$$(e^{2\pi\lambda[A+\varepsilon L(\varepsilon)]} - I)c + \int_0^{2\pi\lambda} e^{(2\pi\lambda - \sigma)A} f[x(\sigma,c,\varepsilon),\varepsilon]d\sigma = 0 \qquad (19)$$

We look for a periodic solution which has the initial value

$$d = (0, s, sd_3(s), \ldots, sd_n(s)) \qquad (20)$$

where s is a real number such that $|s|$ is sufficiently small and $d_3(s)$, $\ldots, d_n(s)$ are continuous functions of s such that

$$d_3(0) = \cdots = d_n(0) = 0$$

Substituting from (20) into (19), we conclude that in order to find a periodic solution it is sufficient to solve the equation:

$$(e^{2\pi\lambda[A+\varepsilon L(\varepsilon)]} - I)d + \int\limits_0^{2\pi\lambda} e^{(2\pi\lambda-\sigma)A} f[x(\sigma,d,\varepsilon),\varepsilon]d\sigma = 0 \qquad (21)$$

for $\varepsilon, \lambda, d_3, \ldots, d_n$ as functions of s. To solve equation (21), we proceed roughly as follows: divide equation (21) by s and apply the implicit function theorem. More precisely, we define the n-vector function $F(s,\varepsilon,\lambda,d_3,\ldots,d_n)$ as follows: if $s \neq 0$,

$$F(s,\varepsilon,\lambda,d_3,\ldots,d_n) = (e^{2\pi\lambda[A+\varepsilon L(\varepsilon)]} - I)\frac{d}{s}$$

$$+ \frac{1}{s}\int\limits_0^{2\pi\lambda} e^{(2\pi\lambda-\sigma)A} f[x(\sigma,d,\varepsilon),\varepsilon]d\sigma$$

If $s = 0$,

$$F(0,\varepsilon,\lambda,d_3,\ldots,d_n) = (e^{2\pi\lambda[A+\varepsilon L(\varepsilon)]} - I)\tilde{d}$$

where

$$\tilde{d} = (0,1,d_3,\ldots,d_n)$$

It is easy to verify that the function $F(s,m\varepsilon,\lambda,d_3,\ldots,d_n)$ is continuous in $(s,\varepsilon,\lambda,d_3,\ldots,d_n)$ for λ in a neighborhood of 1 and $s,\varepsilon,d_3,\ldots,d_n$ in a neighborhood of the 0 in R^n. The only nontrivial part of the verification is proving that

$$\lim_{s\to 0} F(s,\varepsilon,\lambda,d_3,\ldots,d_n) = F(0,\varepsilon,\lambda,d_3,\ldots,d_n) \qquad (22)$$

But this follows easily from the hypothesis that $|f(x,\varepsilon)| = o(|x|)$. (See Exercise 6.) It is easy to verify that $s = 0$, $\varepsilon = 0$, $d_3 = \cdots = d_n = 0$, $\lambda = 1$ is a solution of the equation

$$F(s,\varepsilon,\lambda,d_3,\ldots,d_n) = 0 \qquad (23)$$

Finally, a straightforward computation shows that at $s = 0$, $\varepsilon = 0$, $\lambda = 1$, $d_3 = \cdots = d_n = 0$, the matrix

$$
\begin{bmatrix}
\dfrac{\partial F_1}{\partial \varepsilon} & \dfrac{\partial F_1}{\partial \lambda} & \dfrac{\partial F_1}{\partial d_3} & \cdots & \dfrac{\partial F_1}{\partial d_n} \\[2mm]
\dfrac{\partial F_2}{\partial \varepsilon} & \dfrac{\partial F_2}{\partial \lambda} & \cdot & \cdot\cdot & \dfrac{\partial F_2}{\partial d_n} \\[2mm]
& & \cdot & \cdot & \\
& & \cdot & & \\
\dfrac{\partial F_n}{\partial \varepsilon} & \dfrac{\partial F_n}{\partial \lambda} & \cdot & \cdot & \dfrac{\partial F_n}{\partial d_n}
\end{bmatrix}
\tag{23'}
$$

becomes

$$
\left[
\begin{array}{cc|c}
0 & 2\pi & \\
2\pi & 0 & 0 \\
\hline
& 0 & e^{2\pi D} - I
\end{array}
\right]
\tag{24}
$$

(See Exercise 7.) Since this matrix is nonsingular, then we may solve (23) uniquely for $(\varepsilon, \lambda, d_3, \ldots, d_n)$ in terms of s for $(\varepsilon, \lambda, d_3, \ldots, d_n)$ in a neighborhood of $(0, 1, 0, \ldots, 0)$ and s in a neighborhood of 0. But solving (23) is equivalent to solving equation (21). Summarizing, we have:

Hopf Bifurcation Theorem. *There is an open interval I with midpoint 0 and a neighborhood N in R^n of $(0, 1, 0, \ldots, 0)$ such that for $s \in I$, there is a unique differentiable function*

$$(\varepsilon(s), \lambda(s), d_3(s), \ldots, d_n(s))$$

with range in N such that for each $s \in I$, the solution

$$x(t, d(s), \varepsilon(s))$$

of (18), where

$$d(s) = (0, s, sd_3(s), \ldots, d_n(s))$$

has period $2\pi\lambda(s)$.

Note that this theorem does not guarantee that (18) has periodic solutions for both positive and negative values of ε. For example, the function $\varepsilon(s)$ may have nonnegative values for all $s \in I$.

In Marsden and McCracken [1976], there is an extensive study of the stability of the periodic solutions given by the Hopf Bifurcation Theorem.

A Bifurcation Theorem Based on Sell's Theorem

Bifurcation theorems, especially the Hopf Bifurcation Theorem, have received considerable attention in recent years because they supply models (i.e. descriptions in terms of differential equations) of important biological and chemical phenomena. The hypotheses in the Hopf Bifurcation Theorem are, from the point of view of pure mathematics, quite natural and not excessive. But from the point of view of applications, they may be artificial. Suppose, for example, one tries to describe some aspect of the biological process of morphogenesis (the mechanism of which is not at all well understood) by using a system of differential equations. The condition that an eigenvalue have multiplicity one is not mathematically demanding. But it may have no meaning in terms of the biological concepts being studied. A similar difficulty arises when a chemical process or phenomenon such as the Belousov-Zhabotinsky reaction is studied by means of differential equations.

A simpler result which does not require such stringent hypotheses and which might be useful in modeling biological and chemical phenomena is the following direct consequence of Sell's Theorem:

Theorem 7.10 *Consider the system*

$$x' = f(x, \lambda) \qquad (B)$$

where x is an m-vector, λ is a real parameter and f has continuous first derivatives at each point of the set $U \times I$, where U is an open set in R^n and I is an open interval on R. Suppose that 0 is the only zero of $f(\cdot, \lambda)$ in $\bar{V}$ where $\lambda \in I$ and $\bar{V}$ is the closure of a bounded open set in R^n such that $\bar{V} \subset U$. Suppose further that $\lambda_0 \in I$ is a *fixed value such that every solution of*

$$x' = f(x, \lambda_0)$$

which passes through a point in $\bar{V}$ stays in $\bar{V}$ for all later time. Let

$$L(\lambda) = f_x[0, \lambda]$$

If $L(\lambda_0)$ has a eigenvalue the real part of which is positive and if equation (B) with $\lambda = \lambda_0$ has an asymptotically stable solution which passes through a point in $\bar{V}$ then equation (B) with $\lambda = \lambda_0$ has a nontrivial asymptotically stable periodic solution.

Exercises for Chapter 7

1. Prove: the equation

$$x' = Ax$$

has no nonzero solutions of period T if and only if the eigenvalues of matrix TA are all different from $\pm i2n\pi$ ($n = 0, 1, 2, \ldots$).

2. Use Theorems 7.1 and 7.3 to go back and answer Problem 1. (Include proof that $u(t, \varepsilon) \to u(t)$ as $\varepsilon \to 0$.)

3. Find a real nonsingular matrix H such that

$$H(e^{TA} - I) = P_r$$

4. Prove Theorem 7.6.

5. Investigate the stability of periodic solutions in the resonance case if

$$eTA - I = \begin{bmatrix} 0 & \\ & C_1 \end{bmatrix}$$

6. Prove equation (22) in the proof of the Hopf Bifurcation Theorem.

7. Verify that the matrix $(23')$ in the proof of the Hopf Bifurcation Theorem becomes matrix (24) at $\lambda = 1$, $s = \varepsilon = d_3 = \cdots = d_n = 0$

8. The phenomenon of phase locking occurs in systems of coupled oscillators where the oscillators have, in general, different natural frequencies. If all coupled oscillators begin to oscillate with the same fixed frequency, one says that phase locking takes place. The mathematical problem of phase locking in the case of weak coupling consists in seeking a periodic solution of an autonomous differential equation of the following form.

Let $U = (x_1, \ldots, x_n)$ and $U_j = (x_1^{(j)}, \ldots, x_n^{(j)})$, $j = 1, \ldots, N$, denote points in R^n. We consider the system

$$\frac{dU_1}{dt} = F(U_1) + \varepsilon F_1(U_1, \varepsilon) + \varepsilon G_1(U_1, \ldots, U_N)$$

$$\frac{dU_2}{dt} = F(U_2) + \varepsilon F_2(U_2, \varepsilon) + \varepsilon G_2(U_1, \ldots, U_N) \qquad (*)$$

$$\cdots$$

$$\frac{dU_N}{dt} = F(U_N) + \varepsilon F_N(U_{N,\varepsilon}) + \varepsilon G_N(U_1, \ldots, U_N)$$

where the function

$$F(U) = (F^{(1)}(U), \ldots, F^{(n)}(U))$$

maps R^n into R^n, the functions

$$F_j(U, \varepsilon) = (F_j^{(1)}(U, \varepsilon), \ldots, F_j^{(n)}(U, \varepsilon)) \quad (j = 1, \ldots, N)$$

map $R^n \times I$ into R^n (where I is an interval $(-r, r)$ with $r > 0$ on the real line) and the functions

$$G_j = (G_j^{(1)}, \ldots, G_j^{(n)})$$
$$(j = 1, \ldots, N)$$

map

$$\underbrace{R^n \times R^n \times \cdots \times R^n}_{N \text{ times}}$$

into R^n. For our purposes, it will be sufficient to assume that all functions have continuous third derivatives in all variables. The parameter ε will be such that $|\varepsilon|$ is small.

Each U_j describes an oscillator governed by an equation

$$\frac{dU_j}{dt} = F(U_j) + \varepsilon F_j(U_j, \varepsilon)$$

Since the terms $\varepsilon F_j(U_j, \varepsilon)$ are small, the oscillations U_j are all governed by similar rules or laws. Each εG_j describes the coupling, i.e., the influence of $U_1, \ldots, U_{j-1}, U_{j+1}, \ldots, U_N$ on U_j. Since each εG_j is small, the coupling is said to be *weak*.

We make the following assumptions.

Assumption 1. The equation

$$\frac{dU}{dt} = F(U) \tag{1}$$

has a nontrivial periodic solution $\tilde{U}(t)$ (a limit cycle) of period T and the characteristic exponents of the linear variational equation of (1) relative to $\tilde{U}(t)$, i.e., the equation

$$\frac{dV}{dt} = \{F_U[\tilde{U}(t)]\}V \tag{2}$$

(where F_U denotes the differential of F) are such that $(n-1)$ of the characteristic exponents have negative real parts. (From the existence of the periodic solution $\tilde{U}(t)$, it follows that one characteristic exponent is zero.)

Remark. By Assumption 1, if $\varepsilon = 0$, then $(*)$ has the periodic solution, periodic of period T,

$$\underbrace{(\tilde{U}(t), \ldots, \tilde{U}(t))}_{N \text{ times}}$$

Consequently, it follow from Theorem 7.10 and the Phase Asymptotic Stability Theorem in Chapter 4, that there exist differentiable functions $T_1(\varepsilon), \ldots, T_N(\varepsilon)$, each defined in a neighborhood of zero on the real line, with

$$T_j(0) = T \qquad (j = 1, \ldots, N)$$

such that for $j = 1, \ldots, N$ and for $|\varepsilon|$ sufficiently small the equation

$$\frac{dU_j}{dt} = F(U_j) + \varepsilon F_j(U_j, \varepsilon)$$

has a solution $W_j(t, \varepsilon)$ of period $T_j(\varepsilon)$ and this solution is phase asymptotically stable. $W_j(t, \varepsilon)$ is continuous in (t, ε) and

$$W_j(t, 0) = \tilde{U}(t).$$

The phase-locking problem is:

To show that there exists a function $T(\varepsilon)$ with domain J an open set where

$$0 \in J \subset I$$

and with $T(0) = T$ and such that if $\varepsilon \in J$, then equation $(*)$ has a solution

$$(\mathcal{U}_1(t, \varepsilon), \ldots, \mathcal{U}_N(t, \varepsilon))$$

of period $T(\varepsilon)$ with

$$\lim_{\varepsilon \to 0} \mathcal{U}_j(t, \varepsilon) = \tilde{U}(t) \qquad (j = 1, \ldots, N)$$

for all t (where $\tilde{U}(t)$ is given by Assumption 1).

(Outline of solution: let

$$T(\varepsilon) = T(1 + \varepsilon m)$$

where $m = m(\varepsilon)$ is a differential function of ε and introduce the independent variable:

$$s = (1 + \varepsilon m)^{-1} t$$

The problem then becomes: to determine $m(\varepsilon)$ such that there exists a solution of period of the form

$$(\tilde{U}(s) + \varepsilon X_1(s, \varepsilon), \ldots, \tilde{U}(s), \ldots, X_N(s, \varepsilon))$$

Derive differential equations for $X_1, \ldots, X_N$. Simplify by using Floquet theory and derive a system of $(N - 1)$ bifurcation equations.)

Phase locking occurs in mechanical, electrical, and biological systems. See Andronov and Chaikin [1949] for examples of mechanical and electrical systems. See Winfree [1967] for examples of biological systems. Weakly coupled oscillators occur in electrical systems. Whether phase locking in biological systems can be described by weakly coupled systems is debatable.

Appendix

Ascoli's Theorem

Let $\{f_n\}$ be a sequence of real-valued equicontinuous functions on an interval $[a, b]$, and suppose there exists $M > 0$ such that for all n and for all $x \in [a, b]$,

$$|f_n(x)| < M$$

Then $\{f_n\}$ contains a subsequence which is uniformly convergent on $[a, b]$.

This is the simplest form of Ascoli's Theorem and it is sufficient for our purposes. For a more general version of the theorem, see, for example, Royden [1968, p. 179].

Principle of Contraction Mappings

Definition. Let S be an arbitrary metric space with metric d. A mapping

$$F: S \to S$$

is a *contraction mapping* if there exists a number $k \in (0, 1)$ such that for all $x, y \in S$,

$$d(F(x), F(y)) < kd(x, y)$$

Definition. A metric space S is *complete* if each Cauchy sequence in S has a limit. That is, if $\{x_n\} \subset S$ is such that for $\varepsilon > 0$ there is

321

$$\lim_{n\to\infty} d(x_n, x_0) = 0$$

Principle of Contraction Mappings: If F is a contraction mapping from a complete metric space into itself, then F has a unique fixed point.

Proof. Let x be an arbitrary point of S, and let

$$x_n = F^n(x)$$

where $F^n(x)$ is defined inductively by: $F^1(x) = F(x)$ and $F^n(x) = F[F^{n-1}(x)]$, $n = 2, 3, \dots$. Then $\{x_n\}$ is a Cauchy sequence because

$$d(x_m, x_n) = (F^m(x), F^n(x))$$
$$< k^m d(x, F^{n-m}(x))$$

(where, for definiteness, we assume $n > m$)

$$< k^m \{d(x, x_1) + d(x_1, x_2) + \cdots + d(x^{n-m-1}, x^{n-m})\}$$
$$< k^m \{d(x, x_1) + kd(x, x_1) + \cdots + k^{n-m-1} d(x, x_1)\}$$
$$< k^m \frac{1}{1-k} d(x, x_1)$$

Since $k \in (0, 1)$, then if m is sufficiently large, the quantity

$$\frac{k^m}{1-k} d(x, x_1)$$

can be made arbitrarily small.

Since S is complete, the Cauchy sequence $\{x_n\}$ has a limit x_0. Then x_0 is a fixed point because

$$F(x_0) = F\left[\lim_{n\to\infty} F^n x\right]$$
$$= \lim F^{n+1}(x)$$
$$= \lim_{n\to\infty} x_{n+1}$$
$$= x_0$$

To prove the uniqueness of the fixed point x_0, suppose that there exist two fixed points x and y. Then

$$d(x, y) = d(Fx, Fy) < kd(x, y)$$

Since $k < 1$, then $d(x, y) = 0$ and $x = y$.

Topological Degree

Topological degree is a geometric notion which has proved extremely useful in qualitative studies of various kinds of nonlinear functional equations: ordinary differential equations, partial differential equations, integral equations and integro-differential equations. Although degree theory can be used to obtain only qualitative results of a general nature (existence theorems and some limited results about stability), use of degree theory is an important method in analysis because the results obtained are significant and have not so far been obtained by other methods. Here we will merely sketch the definition of topological degree of mappings in Euclidean spaces (the Brouwer degree), and normed linear spaces (the Leray-Schauder degree) and list those properties of degree which will be useful for the studies in this book.

Our sketch of the definition of degree is based on the definition given by Nagumo [1951]. For a more general and elegant version of the definition, see Nirenberg [1974]. We use the Nagumo definition because it requires the least technical language and is suggestive of methods for computing the degree of particular classes of mappings. Descriptions of the use of topological degree in other studies of functional equations may be found in Cronin [1964], Krasnoselskii [1964], Mawhin and Rouche [1973], Nirenberg [1974]. Definitions of topological degree for more general classes of mappings in normed linear spaces have been given by Nussbaum [1971, 1972, 1974] and Browder and Petryshyn [1969].

Let $\bar{U}$ be the closure of a bounded open set U in R^n and let f be a continuous mapping with domain $\bar{U}$ and range contained in R^n, i.e.

$$f: \bar{U} \to R^n$$

Mapping f can be described by

$$f: (x_1, \ldots, x_n) \to (f_1(x_1, \ldots, x_n), \ldots, f_n(x_1, \ldots, x_n))$$

where $f_1, \ldots, f_n$ are real-valued functions. We will assume that the partial derivatives

$$\frac{\partial f_i}{\partial x_j} \qquad (i,j = 1,\ldots,n)$$

exist and are continuous at each point of U. Let $p \in R^n$ be such that

$$p \notin f(\bar{U} - U)$$

i.e. p is not an image point of the boundary of U. Suppose that $q \in U$ is such that $f(q) = p$. We will say that q is a point of multiplicity $+1$ if

$$\det\left[\frac{\partial f_i}{\partial x_j}(q)\right] > 0$$

i.e. the jacobian of f at q is positive, and q is a point of multiplicity -1 if

$$\det\left[\frac{\partial f_i}{\partial x_j}\right](q) < 0$$

Let us assume that if $q \in f^{-1}(p)$, the Jacobian of f at q is nonzero. It follows easily then that $f^{-1}(p)$ is finite. The topological degree or Brouwer degree of f at p and relative to the set $\bar{U}$, which we denote by $\deg[f, \bar{U}, p]$, is defined to be:

$$\deg[f, \bar{U}, p] = [\text{Number of points } q \text{ in } f^{-1}(p) \text{ of multiplicity } +1]$$
$$- [\text{Number of points } q \text{ in } f^{-1}(p) \text{ of multiplicity } -1]$$

That is, to determine $\deg[f, \bar{U}, p]$ one counts the number of points in $f^{-1}(p)$ that have multiplicity $+1$ and subtracts from it the number of points in $f^{-1}(p)$ that have multiplicity -1.

This definition has two serious deficiencies. First, it is not generally true that the Jacobian of f is nonzero at each point of $f^{-1}(p)$. Indeed, the set $f^{-1}(p)$ is not, in general, finite. Also, it is desirable (for aesthetic and practical reasons) to define $\deg[f, \bar{U}, p]$ for mappings f which are merely continuous rather than restricting our definition to differentiable mappings.

It turns out that if $\bar{f}$ is a continuous mapping from $\bar{U}$ into R^n and if $p \notin \bar{f}(\bar{U} - U)$, then $\bar{f}$ can be approximated arbitrarily closely by a mapping f to which our definition is applicable, i.e. a mapping f

which is differentiable and is such that if $p \notin f(\bar{U}, \bar{U})$, the set $f^{-1}(p)$ is finite, and

$$\det \left[\frac{\partial f_i}{\partial x_j}(q) \right] \neq 0$$

for each $q \in f^{-1}(p)$. Moreover, if $f^{(1)}$ and $f^{(2)}$ are two such mappings which approximate $\bar{f}$ sufficiently well, then

$$\deg \left[f^{(1)}, \bar{U}, p \right] = \deg \left[f^{(2)}, \bar{U}, p \right]$$

Hence if f is a differentiable mapping such that $f^{-1}(p)$ is finite and

$$\det \left| \frac{\partial f_i}{\partial x_j}(q) \right| \neq 0$$

for each $q \in f^{-1}(p)$ and if f approximates $\bar{f}$ sufficiently well, we may define the topological degree or Brouwer degree of $\bar{f}$ at p and relative to $\bar{U}$, denoted by $\deg[\bar{f}, \bar{U}, p]$, as

$$\deg[\bar{f}, \bar{U}, p] = \deg[f, \bar{U}, p]$$

Note that we have omitted entirely the question of: (i) proving that a continuous mapping $\bar{f}$ can be approximated by a differentiable map with the desired properties; (ii) proving that if $f^{(1)}$ and $f^{(2)}$ are two such (sufficiently fine) approximations, then

$$\deg[f^{(1)}, \bar{U}, p] = \deg[f^{(2)}, \bar{U}, p]$$

Such proofs are well-known (Nagumo [1951], Alexandroff and Hopf [1935], Cronin [1964]) but they do not seem to provide any help in applying topological degree to problems in functional equations. Hence, we omit them. The proofs use Sard's Theorem which we state later when we need it for showing how the degree measures the number of solutions.

We add just one remark concerning the geometric meaning of the condition:

$$\det \left[\frac{\partial f_i}{\partial x_j}(q) \right] > 0$$

Since f is differentiable, then for points $\bar{q}$ in a neighborhood of q, the mapping

$$\bar{q} - q \to f(\bar{q}) - f(q)$$

can be approximated by the linear mapping described by the matrix

$$\mathcal{M} = \left[\frac{\partial f_i}{\partial x_j}(q) \right]$$

The condition

$$\det \left[\frac{\partial f_i}{\partial x_j}(q) \right] > 0$$

implies that the mapping described by $\mathcal{M}$ is orientation-preserving. (In the two-dimensional case, if $\mathcal{M}$ is applied to a triangle whose vertices are oriented in counterclockwise order, the images of the vertices will also be oriented in counterclockwise order. A complete description of "orientation-preserving," which we will not give, requires a rigorous definition of "counterclockwise" and an analogous definition in the n-dimensional case where $n > 2$. See Alexandroff and Hopf [1935] and Cronin [1964].)

It is occasionally convenient to use the following special case of the degree. Suppose that mapping f takes $\bar{U}$ into R^n and suppose that $p \in R^n$ is such that $p \notin f(\bar{U} - U)$ and $f^{-1}(p)$ is a single point $q \in U$. From the definition of degree, it follows that if V is any open set in R^n such that $q \in V$ and $\bar{V} \subset \bar{U}$, then

$$\deg[f, \bar{V}, p] = \deg[f, \bar{U}, p]$$

Definition. The *topological index of f at q is*

$$\deg[f, \bar{V}, p]$$

where V is any open set in R^n such that $q \in V$ and $\bar{V} \subset \bar{U}$.

Example 1 If I is the identity mapping from R^n into R^n and U is a bounded open set, then

$$\deg[I, \bar{U}, p] = +1 \quad \text{if} \quad p \in U$$
$$\deg[I, \bar{U}, p] = 0 \quad \text{if} \quad p \in \bar{U}^c$$

This statement follows at once from the definition of Brouwer degree.

Example 2 If A is a linear homogeneous mapping from R^n into R^n and A is described by a nonsingular matrix $\mathcal{M}$, then

$$\deg[A, \bar{U}, p] = \text{sgn det } \mathcal{M} \quad \text{if } p \in A(U)$$
$$\deg[A, \bar{U}, p] = 0 \qquad\qquad \text{if } p \in [\overline{A(U)}]^c$$

This statement also follows at once from the definition of Brouwer degree.

Example 3 If g is a mapping from $\bar{V}$, the closure of a bounded open set V in R^q, into R^1, i.e.

$$g(x_1 \ldots, x_q) \rightarrow (g_1(x_1 \ldots, x_q), \ldots, g_q(x_1, \ldots, x_q))$$

and if f is a mapping from $\bar{U}$, the closure of a bounded open set in R^n, into R^n, where $n > q$, and $\bar{U} \cap R^q = \bar{V}$ and f is defined by

$$f : (x_1, \ldots, x_q, \ldots, x_n)$$
$$\rightarrow (g_1(x_1, \ldots, x_q), \ldots, g_q(x_1, \ldots, x_q), x_{q+1}, \ldots, x_n)$$

Then if

$$p = (x_1^0, \ldots, x_q^0, x_{q+1}^0, \ldots, x_n^0)$$

and $p \notin f(\bar{U} - U)$ and if $p_0 = (x_1^0, \ldots, x_q^0)$ and $p_0 \notin g(\bar{V} - V)$, then

$$\deg[f, U, p] = \deg[g, \bar{V}, p_0]$$

The proof of this result follows from a careful examination of the definition of degree. The mapping f is called a *suspension* of mapping g.

Example 4 If the mapping f is a constant mapping, i.e. for all $q \in \bar{U}$, $f(q) = \tilde{p}$, then if $p \neq \tilde{p}$,

$$\deg[f, \bar{U}, p] = 0$$

Now we describe some of the properties of topological degree which make it useful in analysis. The most fundamental of these properties, which is indeed at the basis of applications, is:

Property 1.   If $\deg[f, \bar{U}, p] \neq 0$, then the set $f^{-1}(p)$ is nonempty. In other words, if $\deg[f, \bar{U}, p] \neq 0$, the equation

$$f(x) = p$$

has a solution $x \in U$.

The proof of Property 1 for differentiable approximating mapping f is trivial. To extend the proof to all continuous f involves only arguments of a routine nature.

Property 1 shows that if it can be proved that

$$\deg[f, \bar{U}, p] \neq 0$$

then we can conclude that the equation

$$f(x) = p$$

has a solution. Thus, the problem of solving an equation can be translated into the problem of computing the degree of a mapping. However, if we look at the definition of $\deg[f, \bar{U}, p]$ and at Property 1, there seems to be a kind of circularity involved. In order to compute $\deg[f, \bar{U}, p]$, we look at the set $f^{-1}(p)$. Then if $\deg[f, \bar{U}, p]$ is shown to be nonzero, we conclude from Property 1 that the equation

$$f(x) = p$$

has a solution, i.e. that the set $f^{-1}(p)$ is nonempty. It seems that in order to establish the mere fact that $f^{-1}(p)$ is nonempty, we must study in some detail the properties of the points in $f^{-1}(p)$! Such a procedure is clearly unreasonable. Actually, we proceed by various indirect methods to show that $\deg[f, \bar{U}, p] \neq 0$ and then invoke Property 1 to conclude that the equation

$$f(x) = p$$

has a solution.

The basis for the most important method for computing $\deg[f, \bar{U}, p]$ is the fact that $\deg[f, \bar{U}, p]$ is constant under a continuous deformation of f or, in more standard terminology, that $\deg[f, \bar{U}, p]$ is invariant under homotopy. The precise statement is:

Property 2. Suppose that F denotes a continuous mapping from $\bar{U} \times [0, 1]$ into R^n. If $t \in [0, 1]$, the mapping $F/\bar{U} \times \{t\}$ will be denoted by F_t. Then the mapping F_t can be identified with the mapping g from $\bar{U}$ into R^n defined by:

$$g: q \rightarrow F_t(q)$$

Let $p(\tau)$ be a continuous curve, i.e. a continuous mapping from the interval $0 \leq \tau \leq 1$ into R^n such that for all $t \in [0, 1]$ and all $\tau \in [0, 1]$,

$$p(\tau) \notin F_t(\bar{U} - U)$$

Then for each $t \in [0, 1]$ and each $\tau \in [0, 1]$, the topological degree

$$\deg[f_t, \bar{U}, p(\tau)]$$

is defined, and for all $t \in [0, 1]$ and $\tau \in [0, 1]$, it has the same value.

The proof of Property 2 is not particularly difficult. But since it seems to shed no light on how to go about applying degree theory in analysis, we omit it.

Definition. The mappings F_0 and F_1 in Property 2, regarded as mappings from $\bar{U}$ into R^n, are said to be *homotopic* in

$$R^n - \{p(\tau) \mid \tau \in [0, 1]\}.$$

Mapping F is a *homotopy* in $R^n - \{p(\tau) \mid \tau \in [0, 1]\}$.

Property 2 is applied in the following way. Suppose we wish to compute $\deg[f, \bar{U}, p]$. We look for a mapping F from $\bar{U} \times [0, 1]$ into R^n such that

(i) $F_0 = f$;

(ii) $p \notin F_t(\bar{U} - U)$ for all $t \in [0, 1]$;

(iii) $\deg[F_1, \bar{U}, p]$ can be easily computed.

It follows from Property 2 that

$$\deg[f, \bar{U}, p] = \deg[F_0, \bar{U}, p] = \deg[F_1, \bar{u}, p]$$

Since we have so far computed the degrees of only a couple of very simple mappings, we have little hope that such a search for a mapping F would be successful. Finding such a mapping F is, in general, very difficult. Nevertheless, the mapping F can be found in a large number of cases which yield significant results in analysis. Now we illustrate the use of Property 2 with some examples.

Example 5 Suppose that $f(\bar{U})$ is contained in an m-dimensional subspace R^m of R^n where $m < n$. Then if $p \in R^n$ is such that $p \notin f(\bar{U} - U)$,

$$\deg[f, \bar{U}, p] = 0$$

Proof. Since $p \notin f(\bar{U} - U)$, there is a connected neighborhood N of p such that

$$N \cap f(\bar{U} - U) = \emptyset$$

But if $\bar{p} \in N$, there is a continuous curve in N which joins p and $\bar{p}$. Hence, by Property 2, if $\bar{p} \in N$, then

$$\deg[f, \bar{U}, p] = \deg[f, \bar{U}, \bar{p}]$$

But there is a point $\tilde{p} \in N$ such that $\tilde{p} \notin R^m$ and hence such that $\tilde{p} \notin f(\bar{U})$. Since $\tilde{p} \notin f(U)$, then by Property 1, it follows that

$$\deg[f, \bar{U}, \tilde{p}] = 0$$

Thus

$$\deg[f, \bar{U}, p] = \deg[f, \bar{U}, \tilde{p}] = 0$$

Example 6 If the mapping

$$f \colon \bar{U} \to R^n$$

is described by

$$f: (x_1, \ldots, x_n) \to (f_1(x_1, \ldots, x_n), \ldots, f_n(x_1, \ldots, x_n))$$

and one of the functions, say $f_j(x_1, \ldots, x_n)$, is nonnegative or non-positive, and if $0 \notin f(\bar{U} - U)$, then

$$\deg[f, \bar{U}, 0] = 0$$

Proof. If $f_j(x_1, \ldots, x_n)$ is nonnegative, then if $(\bar{x}_1, \ldots, \bar{x}_n)$ is such that $\bar{x}_j < 0$,

$$(\bar{x}_1, \ldots, \bar{x}_n) \notin f(\bar{U})$$

The remainder of the proof is very similar to that for Example 5.

Example 7 Let f be the mapping from R^2 into R^2 described in terms of a complex variable by:

$$f: z \to z^n$$

where n is a positive integer. Let U be a bounded open set in R^2 such that $0 \in U$. Then

$$\deg[f, \bar{U}, 0] = n$$

Proof. Let $z_0 \neq 0$ be sufficiently close to 0 so that $f^{-1}(z_0) \subset U$. From Property 2, it follows that

$$\deg[f, \bar{U}, 0] = \deg[f, \bar{U}, z_0]$$

(To apply Property 2, let the set $\{p(\tau) \mid \tau \in [0, 1]\}$ be the line segment joining 0 and z_0.) To compute $\deg[f, U, z_0]$, note that $f^{-1}(z_0)$ is the set of n-th roots of z_0 which is a set of n distinct complex numbers $w_1, \ldots, w_n$ each of which is nonzero. To find the sign of the Jacobian of f at these points, we notice that if

$$f(z) = u(x, y) + iv(x, y)$$

then the determinant whose sign must be found is

$$\det \begin{bmatrix} u_x & v_x \\ u_y & v_y \end{bmatrix} = u_x v_y - v_x u_y$$

But by the Cauchy-Riemann equations (which certainly hold for the simple analytic function $f(z) = z^n$)

$$u_x v_y - v_x u_y = u_x^2 + v_x^2$$

But

$$u_x^2 + v_x^2 = |f'(x)|^2 = |nz^{n-1}|^2$$

and for $j = 1, \ldots, n$

$$|nw_i^{n-1}|^2 > 0$$

Hence at each point $w_j (j = 1, \ldots, n)$ the Jacobian is positive, which completes the proof. (The kind of argument used in this example is applicable to a wide class of mappings as we will show later.)

We prove now a generalization of a well-known theorem in complex variable.

Rouché's Theorem *Let $\bar{U}$ be the closure of a bounded open set U in R^n, and let f and g be continuous mappings from $\bar{U}$ into R^n. Suppose that $p \in f(\bar{U} - U)$ and that for all $x \in \bar{U} - U$,*

$$|g(x) - f(x)| < |f(x) - p|$$

Then

$$\deg[g, \bar{U}, p] = \deg[f, \bar{U}, p]$$

Proof. Since $p \notin f(\bar{U} - U)$, then $\deg[f, \bar{U}, p]$ is defined. The mapping

$$f(x) + t[g(x) - f(x)]$$

is a homotopy in $R^n - \{p\}$ because if $x \in \bar{U} - U$,

$$|f(x) + t[g(x) - f(x)] - p|$$
$$\geq |f(x) - p| - t|g(x) - f(x)|$$
$$\geq |f(x) - p| - |g(x) - f(x)| > 0$$

Hence f and g are homotopic in $R^n - \{p\}$ and by Property 2, the conclusion of the theorem holds. □

Example 8 Suppose mapping f is defined on $\bar{U}$ where U is an open neighborhood of the origin 0 and f has the form

$$f(x) = \mathcal{M}x + h(x)$$

where $\mathcal{M}$ is a nonsingular matrix and

$$h(x) = o(x)$$

i.e.,

$$|h(x)| < \eta(|x|)|x|$$

where $\eta(r)$ is a continuous monotonic increasing function from the set of nonnegative real numbers into the nonnegative real numbers such that

$$\eta(0) = 0$$

Then there exists a number $r_0 > 0$ such that if V is an open neighborhood of the origin such that $V \subset B_{r_0}$, the ball of radius r_0 and center 0, then

$$\deg[f, \bar{V}, 0] = \text{sgn det } \mathcal{M}$$

Proof. Since $\mathcal{M}$ is nonsingular, there is a positive number m such that if $|x| \neq 0$, then

$$|\mathcal{M}x| > m|x|$$

Let r_0 be such that $\eta(r_0) < m$. Then if $0 < |x| < r_0$

$$|f(x) - \mathcal{M}x| = |\mathcal{M}x + h(x) - \mathcal{M}x| = |h(x)| < \eta(r_0)|x| < m|x| < |\mathcal{M}x|$$

By Example 2

$$\deg[\mathcal{M}, \bar{V}, 0] = \text{sgn det } \mathcal{M}$$

Hence the conclusion follows from Rouché's Theorem.

Example 9 Suppose mapping f is defined on R^n and f has the form

$$f(x) = \mathcal{M}x + k(x)$$

where $\mathcal{M}$ is a nonsingular matrix and

$$|k(x)| < N(|x|)|x|$$

where $N(r)$ is a continuous function from the set of positive numbers into the positive numbers and

$$\varlimsup_{r \to \infty} N(r) < m$$

where m has the same meaning as in Example 8. Then there exists a number $r_1 > 0$ such that if W is an open neighborhood of 0 such that $B_{r_1} \subset W$, then

$$\deg[f, \bar{W}, 0] = \operatorname{sgn} \det \mathcal{M}$$

Proof. Let r_1 be such that if $r \geq r_1$, then $N(r) < m$. It follows that if $x \in \bar{W} - W$,

$$|f(x) - \mathcal{M}x| = |\mathcal{M}x + k(x) - \mathcal{M}x| = |k(x)| < N(r_1)|x| < m|x| < |\mathcal{M}x|$$

Again the conclusion follows from Rouché's Theorem.

Another property of topological degree which is useful for computing the degree is the product theorem, i.e. the statement which says roughly that the product of the degrees is the degree of the product. The precise statement is:

Product Theorem *Let $\bar{U}$ be the closure of a bounded open set in R^n and f a continuous mapping from $\bar{U}$ into R^n. Let g be a continuous mapping from $f[\bar{U}]$ into R^n and suppose $p \in R^n$ is such that*

$$p \notin gf(\bar{U} - U)$$

Assume further that $g^{-1}(p)$ is a finite set of points $q_1, \ldots, q_m$ such that for $j = 1, \ldots, m$,

$$q_j \in Int\, f[U]$$

Let $V_1, \ldots, V_m$ be a collection of pairwise disjoint open sets such that for $j = 1, \ldots, m$,

$$q_j \in V_j$$

and

$$\bar{V}_j \subset f[U]$$

Then

$$\deg[gf, \bar{U}, p] = \sum_{j=1}^{n} \deg[f, \bar{U}, q_j] \deg[g, \bar{V}_j, p]$$

The proof of this theorem is not difficult and can be based on a careful examination of the definition of degree.

Example 10 Let f be a mapping from R^{2m} into R^{2m}, where $m > 1$, defined as follows. Regard R^{2m} as complex Euclidean m-space and let f be defined by:

$$f : (z_1, \ldots, z_m) \rightarrow (z_1^{k_1}, \ldots, z_m^{k_m})$$

Then f is the product of the mappings

$$f_j : (z_1, \ldots, z_m) \rightarrow \left(z_1, \ldots, z_j^{k_j}, \ldots, z_m \right) \quad (j = 1, \ldots, m)$$

Hence by Example 3, Example 7 and the Product Theorem, if B^{2m} is a ball in R^{2m} with center 0, then

$$\deg[f, B^{2m}, 0] = \prod_{j=1}^{m} k_j$$

(We note that $\deg[f, B^{2m}, 0]$ can also be computed directly from the definition of topological degree by using the technique used in Example 7. Showing that the Jacobians are positive is rather tedious.)

Example 11 Let f be a mapping from R^{2m} into R^{2m}, where $m > 1$, defined as follows. Regard R^{2m} as complex Euclidean m-space and let f be defined by:

$$f: (z_1, \ldots, z_m) \to (f_1(z_1, \ldots, z_m), \ldots, f_m(z_1, \ldots, z_m))$$

where $f_j(z_1, \ldots, z_m)$ is a polynomial homogeneous of degree k_j in $z_1, \ldots, z_m$ for $j = 1, \ldots, m$. if $R(f_1, \ldots, f_m)$ is the resultant of the polynomials $f_1, \ldots, f_m$, then a classical theorem (see Macaulay [1916] or van der Waerden [1940]) states that $R(f_1, \ldots, f_m) = 0$ if and only if the equations

$$f_1(z_1, \ldots, z_m) = 0$$
$$f_m(z_1, \ldots, z_m) = 0$$

have a nonzero solution. In the language of degree theory, we may state this theorem as: if B^{2m} is the ball with radius one and center 0, in R^{2m}, then $\deg[f, B^{2m}, 0]$ is defined if and only if $R(f_1, \ldots, f_m) \neq 0$. Suppose $R(f_1, \ldots, f_m) \neq 0$. Then

$$\deg[f, B^{2m}, 0] = \prod_{j=1}^{m} k_j$$

Proof. See Cronin [1964].

Example 12 Let f be a mapping from R^{2m} into R^{2m}, where $m > 1$, defined by:

$$f: (z_1, \ldots, z_m) \to (f_1(z_1, \ldots, z_m)$$
$$+ h_1(z_1, \ldots, z_m), \ldots, f_m(z_1, \ldots, z_m) + h_m(z_1, \ldots, z_m))$$

where $f_j(z_1, \ldots, z_m)$ is a polynomial homogeneous of degree k_j in $z_1, \ldots, z_m$, $h_j(z_1, \ldots, z_m)$ is a continuous function of order $k_j + 1$, i.e. if $r^2 = \sum_{j=1}^{m} |z_j|^2$, then

$$\lim_{r \to 0} \frac{|h_j(z_1, \ldots, z_m)|}{r^{k_j}} = 0$$

If the resultant $R(f_1 m, \ldots, f_m)$ is nonzero then there is a positive number b such that if B^{2m} is a ball with center 0 and radius less than b then

$$\deg[f, B^{2m}, 0] = \prod_{j=1}^{m} k_j$$

If $R(f_1, \ldots, f_m) = 0$ and if $\deg[f, B^{2m}, 0]$ is defined and if the radius of B^{2m} is less than some positive number b, then

$$\deg[f, B^{2m}, 0] \geq \prod_{j=1}^{m} k_j$$

Proof. For the case in which $R(f_1, \ldots, f_m) \neq 0$, the proof is obtained by a straightforward application of Rouché's Theorem. For the proof of the second case, see Cronin [1964].

Example 13 Let f be defined by:

$$f: (z_1, \ldots, z_m) \rightarrow (f_1(z_1, \ldots, z_m)$$
$$+ s_1(z_1, \ldots, z_m), \ldots, f_m(z_1, \ldots, z_m) + s_m(z_1, \ldots, z_m))$$

where $f_1, \ldots, f_m$ are as in Example 12, and $R(f_1, \ldots, f_m) \neq 0$ and $s_j(z_1, \ldots, z_m)$ is a continuous function of order less than k_j, i.e. if $r^2 = \sum_{j=1}^{m} |z_j|^2$, then

$$\lim_{r \to \infty} \frac{|s_j(z_1, \ldots, z_m)|}{r^{k_j}} = 0$$

Then there is a positive number b_1 such that if B^{2m} is a ball with center 0 and radius greater than b_1, then

$$\deg[f, B^{2m}, 0] = \prod_{j=1}^{m} k_j$$

Proof. By an argument similar to that for Example 12.

Example 14 Let h be a homeomorphism from $\bar{U}$, the closure of a bounded connected open set U in R^n, into R^n. Then if $p \in h(U)$,

$$\deg[h, \bar{U}, p] = \pm 1$$

Proof. Apply the Product Theorem with $f = h$ and $g = h^{-1}$, and use the fact that the topological degree of the identity mapping at points in the image set is $+1$ (Example 1). Note that in order to be certain that $\deg[h^{-1}, h(\bar{U}), h^{-1}(p)]$ is defined, one must know that $h(\bar{U})$ is the closure of a bounded open set. This follows from the Invariance of Domain Theorem (see, e.g., Hurewicz and Wallman [1948]). At the cost of extra labor, the Invariance of Domain Theorem itself can be proved by using the Product Theorem. See Leray [1935].

The definition of topological degree that we have described is the "covering number" definition which was given originally by Brouwer [1912]. From the point of view of the analyst who is interested in solving equations, it is a very natural definition. There is, however, an equivalent definition which is equally important. The second definition, which is sometimes called the "intersection number" definition, is obtained as follows. Suppose, as before, that $\bar{U}$ is the closure of a bounded open set $U \subset R^n$, that f is a continuous mapping from $\bar{U}$ into R^n and that

$$p \notin f(\bar{U} - U)$$

Let L be a half-ray in R^n emanating from the point p, i.e.

$$L = \{p + \lambda \vec{v} \mid \lambda \geq 0\}$$

where $\vec{v}$ is a fixed nonzero n-vector. Let K be the set of points of intersection of L and $f(\bar{U} - U)$, i.e.

$$K = L \cap \{f(\bar{U} - U)\}$$

Suppose that $p_i \in K$ and $q_i = f^{-1}(p_i)$. By standard but tedious considerations of orientation, an intersection multiplicity of $+1$ or -1 can be assigned to the point p_i. This is illustrated for the case $n = 2$ in Figure 1. (The "counterclockwise orientation" in R^2 is assumed to be the positive orientation.) The point p_1 has intersection multiplicity $+1$ because r_1, r_2, q_1 and $f(r_1)$, $f(r_2)$, p_1 both have the same orientation. The point p_2 has intersection multiplicity -1 because s_1, s_2, q_2 and $f(s_1)$, $f(s_2)$, p_2 have opposite orientations.

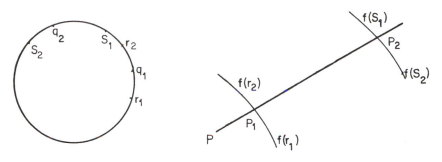

Figure 1

Then $\deg[f,\bar{U},p]$ is defined to be:

$\deg[f,\bar{U},p]$

$= [\text{Number of points in } K \text{ with intersection multiplicity } +1]$

$- [\text{Number of points in } K \text{ with intersection multiplicity } -1]$

As with our presentation of the covering number definition, this definition has many "holes" in it. In the first place, we have only indicated the "orientation considerations" that must be made. Secondly, we have not indicated how the half-ray L should be chosen. A fairly lengthy investigation shows that the definition is independent of the half-ray L. "Well-behaved" approximations g of f are used and it is shown that for sufficiently fine approximations g, the value of $\deg[g,\bar{U},p]$ is always the same.

At the cost of considerable labor, it can be shown that the covering number definition and the winding number definition just described are equivalent. A detailed discussion is given in Alexandroff and Hopf [1935] and Cronin [1964]. Notice that the fact that the two definitions are equivalent implies that the topological degree depends only on the values of f on $\bar{U} - U$, i.e. on the function $f/(\bar{U} - U)$. Indeed the function f need only be defined on $\bar{U} - U$.

There are two important reasons for introducing the intersection number definition. First, the intersection number definition contains as a special case (the case in which $n = 2$) the concept of the winding number, familiar from complex variable. Secondly, use of the intersection number definition of topological degree makes possible the computation of the topological degree of some further classes of mappings as we now show.

Example 15 Let

$$f: (x, y) \to (\bar{x}, \bar{y})$$

be a mapping from R^2 into R^2 defined by:

$$\bar{x} = P(x, y)$$
$$\bar{y} = Q(x, y)$$

where P, Q are polynomials homogeneous in x and y of degrees p and q respectively. Let

$$B^2 = \{(x, y) \mid x^2 + y^2 \leq 1\}$$

We assume that $\deg[f, B^2, 0]$ is defined, i.e. we assume that there is no point $p \in B^2 - \{\text{Int } B^2\}$ such that

$$f(p) = 0$$

This means we assume that P and Q have no common real linear factors. Now we write P and Q in factored form, i.e.

$$P(x, y) = K_1 \prod_{k=1}^{m} (y - a_i x)^{p_i}$$
$$Q(x, y) = K_2 \prod_{j=1}^{n} (y - b_j x)^{q_j}$$

where K_1, K_2 are constants. (We include the possibility that some $a_i = \infty$ or some $b_j = \infty$; equivalently, that the factor $y - a_i x$ is equal to $-x$ or the factor $y - b_j x$ is equal to $-x$.) It follows from the definition of topological degree that if

$$g: (x, y) \to (\tilde{x}, \tilde{y})$$

is the mapping from R^2 into R^2 defined by

$$\tilde{x} = \prod_{i=1}^{n}(y - a_i x)^{p_i}$$

$$\tilde{y} = \prod_{j=1}^{n}(y - b_j x)^{q_j} \tag{1}$$

then

$$\deg[g, B^2, 0] = (\deg[f, B^2, 0])(\text{sign } K_1 K_2)$$

In order to compute $\deg[g, B^2, 0]$ we use the intersection number definition and investigate the image under g of $B^2 - (\text{Int } B^2)$ by studying the changes of sign of $\tilde{x}$ and $\tilde{y}$ as (x, y) varies over $B^2 -$ (Int B^2). By using Property 2 (the invariance under homotopy of the degree) it can be easily shown (Cronin [1964, pp. 39-40]) that the value of $\deg[g, B^2, 0]$ is not affected if the following factors in the products on the right-hand side of (1) above are omitted.

(1) Pairs of factors $(y - a_{i_1}x)$, $(y - a_{i_2}x)$ in which a_{i_1} and a_{i_2} are complex conjugates.

(2) Factors $(y - a_{i_1}x)^{p_i}$ wehre a_{i_1} is real and p_i is even.

(3) Pairs of factors $(y - a_{i_1}x)$, $(y - a_{i_2}x)$ such that $a_{i_1} < a_{i_2}$ and there is no b_j or a_{i_3} such that

$$a_{i_1} < b_j < a_{i_2}$$

or

$$a_{i_1} < a_{i_3} < a_{i_2}$$

(4) Pairs of factors $(y - \tilde{a}_1 x)$, $(y - \tilde{a}_m x)$ where $\tilde{a}_1$ and $\tilde{a}_m$ are the smallest and largest of all the numbers $a_1, \ldots, a_m, b_1, \ldots, b_n$.

(5) The factors listed above with the a_i's replaced by b_j's and the b_j replaced by a_i.

If all the factors in

$$\prod_{i=1}^{m}(y - a_i x)^{p_i}$$

and in

$$\prod_{j=1}^{n}(y - b_j x)^{q_j}$$

are included in the above four classifications, then $\deg[g, B_1, 0]$ is zero because g is homotopic in $R^n - \{0\}$ to a mapping

$$M : (x, y) \rightarrow (x_1, y_1)$$

where x_1 or y_1 is a constant. To complete the study we need only consider a mapping M_1 such that

$$a_1 < b_1 < a_2 < \cdots < a_m < b_m \tag{2}$$

or

$$b_1 < a_1 < b_2 < \cdots < b_m < a_m \tag{3}$$

and with each factor $(y - a_i x)$ and $(y - b_j x)$ having exponent one. It is easily shown (Cronin [1964, p. 40]) by using the intersection number definition of topological degree that if (2) holds, then $\deg[g, B_1, 0]$ is m. If (3) holds, then $\deg[g, B_q, 0]$ is $-m$.

Property 1 states that if the degree of the mapping is nonzero, then the corresponding equation has at least one solution. Since the degree is an integer, it is natural to raise the question of whether the degree or its absolute value measures or counts the number of solutions of the equation. The answer to this question is not obvious. By Example 7, the degree of the mapping

$$f : z \rightarrow z^n$$

relative to the closed unit disc and at 0 is n. But the equation

$$f(z) = 0$$

has just one solution $z = 0$. So the degree (or its absolute value) is not, in general, a lower bound for the number of solutions. On the other hand, suppose the mapping h with domain

$$B^2 = \{(x,y) \mid x^2 + y^2 \leq 1\} = \{(r,\theta) \mid 0 \leq r \leq 1\}$$

is defined by

$$h(x,y) = (\lambda(r)x, \lambda(r)y)$$

where

$$\lambda(r) = 0 \quad \text{if} \quad 0 \leq r \leq \frac{1}{2}$$

$$\lambda(r) = 2\left(r - \frac{1}{2}\right) \quad \text{if} \quad r \geq \frac{1}{2}$$

Mapping h is the identity mapping on the circle $r = 1$. Hence

$$\deg[h, B^2, 0] = +1$$

But the equation

$$h(r, \theta) = (0, 0)$$

has an infinite set of solutions, i.e. the set

$$\left\{(r,\theta) \mid 0 \leq r \leq \frac{1}{2}\right\}$$

So the degree (on its absolute value) is certainly not an upper bound for number of solutions.

However, we can show that if the mapping f is differentiable, then the degree does yield a kind of count of the number of solutions. To prove this we need a special case of a well-known result in analysis.

Definition. Let f be a differentiable mapping of an open set $G \subset R^n$ into R^n. A point $x \in G$ at which the Jacobian of f is zero is a *critical point* of f.

Sard's Theorem (Special case) *Let f be a differentiable mapping of a bounded open set $G \subset R^n$ into R^n. Let D be an open set such that $\bar{D} \subset G$. Then the image under f of the set of critical points in D has measure zero.*

For a proof of Sard's Theorem, see Sard [1942] or Golubitsky and Guillemin [1973].

Sard's Theorem yields almost immediately the following theorem which shows that if f is differentiable, then the absolute value of the degree of f is essentially a lower bound for the number of solutions of the corresponding equation.

Theorem 1 *If f is a differentiable mapping of a bounded open set $G \subset R^n$ into R^n and if $G \supset \bar{U}$, where $\bar{U}$ is the closure of a bounded open set U, and if $\deg[f, \bar{U}, p]$ exists and $\deg[f, \bar{U}, p] = m$, then there is an open neighborhood N of p and a set $S \subset N$ such that S has n-measure zero and such that:*

(i) *For all $\bar{p} \in N$,*

$$\deg[f, \bar{U}, \bar{p}] = \deg[f, \bar{U}, p];$$

(ii) *if $\bar{p} \in N - S$, then $[f^{-1}(\bar{p})] \cap U$ is a finite set of points;*

(iii) *if $\bar{p} \in N - S$ and $q \in [f^{-1}(\bar{p})] \cap U$, then the Jacobian of f at q is nonzero.*

(iv) *if $\bar{p} \in N - S$ and n is the number of points in $[f^{-1}(\bar{p})] \cap U$, then $n \geq |m|$.*

(Sard's Theorem is used as a basis for one way of setting up the definition of topological degree. See Nagumo [1951].)

Mappings defined by analytic functions have the useful property that the degree of the mapping is positive if and only if the corresponding equation

$$f(z) = p$$

has a solution. Moreover the degree is equal to the number of solutions of the equation

$$f(z) = \bar{p}$$

for all $\bar{p}$ in a neighborhood of p except for a finite set. The underlying reason for this is quite simple. Suppose that

$$f(z) = u(x,y) + iv(x,y)$$

is a function analytic in an open set $\mathcal{U}$ in the complex plane. We consider the corresponding mapping

$$f: (x,y) \rightarrow (u(x,y), v(x,y))$$

Suppose that

$$f(x,y) = 0$$

has a solution $(\bar{x}, \bar{y})$, i.e.

$$u(\bar{x}, \bar{y}) = 0$$
$$v(\bar{x}, \bar{y}) = 0$$

From the Cauchy-Riemann equations, it follows that the Jacobian of f at $(\bar{x}, \bar{y})$ is

$$\det \begin{bmatrix} u_x & u_y \\ v_x & v_y \end{bmatrix} = \det \begin{bmatrix} u_x & -v_x \\ v_x & u_x \end{bmatrix} = u_x^2 + v_x^2 \tag{4}$$

Since $\partial f/\partial z$ is an analytic function and

$$\frac{\partial f}{\partial z} = u_x(x,y) + iv_x(x,y)$$

then from the Identity Theorem for Analytic Functions, it follows that in a bounded set in R^2, there is only a finite set of points $(\bar{x}_1, \bar{y}_1), \ldots, (\bar{x}_m, \bar{y}_m)$ such that

$$u_x(\bar{x}_j, \bar{y}_j) + iv_x(\bar{x}_j, \bar{y}_j) = 0 \qquad (j = 1, \ldots, m)$$

Hence if $\bar{U}$ is the closure of a bounded open set in R^2 such that $\bar{U} \subset \mathcal{U}$, then if

$$p \notin f\left[\left\{ \sum_{j=1}^{m} (\bar{x}_j, \bar{y}_j) \right\} \cup [\bar{U} - U] \right]$$

it follows from (4) that $\deg[f, \bar{U}, p]$ is nonnegative and is equal to the number of solutions in the set U of the equation

$$f(z) = p$$

If C^n denotes complex Euclidean n-space, extension of the reasoning above and use of Theorem 1 yields the following result.

Theorem 2 *Suppose that*

$$f_j(z_1, \ldots, z_n) \qquad (j = 1, \ldots, n)$$

is analytic (i.e. can be represented by a power series in $z_1, \ldots, z_n$) in an open set $\mathcal{U}$ in C^n. Let $\bar{U}$ be the closure of a bounded open set U in R^{2n} (where R^{2n} is identified with C^n) such that $\bar{U} \subset \mathcal{U}$. Let f be the mapping of R^{2n} into R^{2n} described in terms of points in C^n by

$$f: (z_1, \ldots, z_n) \rightarrow (f_1(z_1, \ldots, z_n), \ldots, f_n(z_1 \ldots, z_n))$$

Then if $p \notin f(\bar{U} - U)$ and $p \in f(U)$,

$$\deg[f, \bar{U}, p] > 0$$

Also there is a set $S \subset R^{2n}$ such that S has $2n$-measure zero and such that if

$$p \in f(U) - S$$

the number of solutions of the equation

$$f(z) = p$$

is equal to $\deg[f, \bar{U}, p]$.

Further considerations of the same kind yield the following theorem.

Theorem 3 *Suppose that*

$$f_j(z_1, \ldots, z_n) \qquad (j = 1, \ldots, n)$$

is analytic (i.e. can be represented by a power series in $z_1, \ldots, z_n$) in an open set $\mathcal{U}$ in C^n and suppose further that

$$f_j(\bar{z}_1, \ldots, \bar{z}_n) = \overline{f_j(z_1, \ldots, z_n)} \tag{5}$$

where $\bar{z}_i$ denotes the conjugate of z_i (condition (5) will be satisfied, for example, if the coefficients in the power series expansion of f_j $(j = 1, \ldots, n)$ are real.) Suppose $\bar{U}$ is the closure of a bounded open set U in R^n such that $\bar{U} \subset \mathcal{U}$ where R^n is identified with the subspace of the points

$$(z_1, \ldots, z_n) = (x_1 + iy_1, \ldots, x_n + iy_n)$$

in C^n that consists of points of the form $(x_1, \ldots, x_n)$. Let h be the mapping from R^n into R^n defined by

$$h: (x_1, \ldots, x_n) \rightarrow (f_1(x_1, \ldots, x_n), \ldots, f_n(x_1, \ldots, x_n))$$

Suppose further that $\mathcal{V}$ is an open set in C^n such that $\bar{\mathcal{V}} \subset \mathcal{U}$ and $\mathcal{V} \cap R^n = U$. Then there is a set $M \subset R^n$ of n-measure zero such that if

$$p \in R^n - M - h(\bar{U} - U)$$

(so that $\deg[f, \bar{\mathcal{V}}, p]$ is defined) and

$$\deg[h, \bar{U}, p] \neq 0$$

then the number n_h of solutions of the equation

$$h(x) = p$$

is finite and

$$n_h \leq \deg[f, \bar{\mathcal{V}}, p]$$

and

$$n_h = \deg[f, \bar{\mathcal{V}}, p] \pmod 2$$

Proof. See Cronin [1960, 1971].

Example 16 Let f be a mapping from C^n (or R^{2n}) into C^n (or R^{2n}) defined by

$$f: (z_1, \ldots, z_n) = (f_1(z_1, \ldots, z_n), \ldots, f_n(z_1, \ldots, z_n))$$

where for $j = 1, \ldots, n$,

$$f_j(z_1, \ldots, z_n) = P_{k_j}(z_1, \ldots, z_n) + H_{k_j}(z_1, \ldots, z_n)$$

where P_{k_j} is a polynomial homogeneous of degree k_j in $z_1, \ldots, z_n$ and $H_{k_j}(z_1, \ldots, z_n)$ is of order $k_j + 1$ in the sense that if

$$r^2 = \sum_{i=1}^n |z_i|^2$$

$$\lim_{r \to 0} \frac{|H_{k_j}(z_1, \ldots, z_n)|}{r^{k_j}} = 0$$

If B_r is a ball of radius r and center 0 in R^{2n} and if r is sufficiently small, then

$$\deg[f, B_r, 0] \geq \prod_{j=1}^n k_j \qquad (5')$$

Proof. This is proved by using Sard's Theorem. See Cronin [1953].

We have already seen numerous examples of mappings with degree zero. It is natural to raise the question of whether the fact that the degree is zero yields information about the existence of solutions. First, if the degree is zero, the corresponding equation may nevertheless have solutions. By Example 6, the mapping

$$f: (x, y) \to (x^2, y^2)$$

on the unit disc B^2 is such that

$$\deg[f, B^2, 0] = 0$$

But the equation

$$f(x, y) = (0, 0)$$

has the solution: $x = 0$, $y = 0$. Thus the fact that the degree is zero seems an inconclusive observation. However, this fact can be utilized in some cases to obtain useful results. In order to see how it is used, we state one more basic property of the topological degree.

Property 3. Suppose that f is a continuous mapping from $\bar{U} \cup \bar{V}$, where U and V are bounded open sets in R^n and $U \cap V = \emptyset$, into R^n. Let $p \in R^n$ be such that

$$p \notin [f(\bar{U} - U)] \cup [f(\bar{V} - V)]$$

Then

$$\deg[f, \bar{U}, p] + \deg[f, \bar{V}, p] = \deg[f, \overline{U \cup V}, p]$$

Property 3 follows easily from the covering number definition of degree and we will not give a detailed proof.

Using Property 3, we have immediately the following theorem.

Theorem 4 *Suppose U is a bounded open set in R^n and W is an open set such that $W \subset U$. If f is a mapping from $\bar{U}$ into R^n and*

$$p \in R^n - f(\bar{W} - W) - f(\bar{U} - U)$$

and if

$$\deg[f, \bar{W}, p] \neq \deg[f, \bar{U}, p] \tag{6}$$

then the equation

$$f(x) = p$$

has a solution $x \in U - \bar{W}$.

Proof. By Property 3,

$$\deg[f, \bar{W}, p] + \deg[f, \overline{U - W}, p] = \deg[f, \bar{U}, p]$$

Hence by (6),

$$\deg[f, \overline{U - W}, p] \neq 0$$

and the theorem follows from Property 1.

Theorem 4 does not seem a very impressive statement, but it is frequently useful in analysis. It is often possible to show that

$$\deg[f, \bar{U}, p] = 0$$

or $\deg[f, \bar{U}, p]$ is even and to show that f is a homeomorphism in an open neighborhood $\mathcal{N}$ of some point in $f^{-1}(p)$ or that the differential of f at some point in $f^{-1}(p)$ is nonsingular. Then by Example 14, $\deg[f, \mathcal{N}, p] = \pm 1$ and the hypotheses of Theorem 4 are satisfied.

For work in differential equations, it is often convenient to think in terms of a vector field. Analytically, the concept of vector field is, in a simple way, equivalent to the concept of a mapping.

Definition. Let E be a subset of R^n. A *vector field* on E, denoted by $V(x)$ where $x \in E$, is a mapping from E into the collection of real n-vectors. (Geometrically, one thinks of the vector $V(x)$ attached to the point x so that the initial point of $V(x)$ is the point x.) From the analytic viewpoint, the vector field $V(x)$ is simply a mapping M_v from E into R^n, i.e. the mapping from x into the head of the vector $V(x)$ if the initial point of $V(x)$ is placed at the origin. We say that the vector field $V(x)$ is *continuous* or *differentiable* if M_v is continuous or differentiable.

Notice that an n-dimensional autonomous system

$$x' = f(x)$$

can be regarded geometrically simply as a vector field $f(x)$. We make use of this viewpoint only briefly (Exercise 4 in Chapter 6) but it is of basic importance in some studies of differential equations.

Definition. A *singularity* of vector field $V(x)$ is a point $\bar{x} \in E$ such that $V(\bar{x}) = 0$. If vector field $V(x)$ has no singularities on E, then $V(x)$ is *nonsingular*.

Definition. Suppose that $E = \bar{U}$ where $\bar{U}$ is the closure of a bounded open set U in R^n, and suppose that if $y \in \bar{U} - U$, then $V(y) \neq 0$. The *index* of $V(x)$ is $\deg[M_v, \bar{U}, 0]$.

Definition. Suppose $E = \bar{U}$ and $p \in U$ is such that $V(p) = 0$ and there is a neighborhood N of p such that $V(x)$ is nonzero at each point $x \in \bar{N} - \{p\}$. The *index of the vector field $V(x)$ at the singularity p* is the topological index of M_v at p.

Now let $\bar{U}$ be the closure of a bounded open set U in R^n such that at each point q of the boundary of U there is a tangent hyperplane H_q. This will hold if, for example, there is a neighborhood N of q in $\bar{U} - U$ such that N is the image of an open set in R^{n-1} under a one-to-one mapping of the form

$$(x_1, \ldots, x_{n-1}) \rightarrow (f_1(x_1, \ldots, x_{n-1}), \ldots, f_{n-1}(x_1, \ldots, x_{n-1}))$$

where each f_j $(j = 1, \ldots, n-1)$ is a differentiable function. (In the language of differential geometry, the condition holds if $\bar{U} - U$ is a differentiable manifold.) Then if L_q is the line through q which is normal to H_q, there exists a fixed n-vector V_q of unit length such that

$$L_q = \{p(\lambda) \mid p(\lambda) = q + \lambda V_q, \ \lambda \text{ real}\}$$

and such that there exists a number $\eta > 0$ with the property that if $\lambda \in (0, \eta)$, then $p(\lambda) \in U$ and if $\lambda \in (-\eta, 0)$ then $p(\lambda) \in \bar{U}^c$, the complement of $\bar{U}$ relative to R^n.

Theorem 5 *Let $V(x)$ be a nonsingular vector field on $\bar{U} - U$ such that for each $q \in \bar{U} - U$, the vector $V(q)$ is such that*

$$(V(q), V_q) \geq 0 \tag{10}$$

Then the index of $V(x)$ is equal to the index of the vector field V_x.

Proof. The index of $V(x)$ is, by definition, $\deg[M_v, \bar{U}, 0]$. (Remember that the degree depends only on $M_v/(\bar{U} - U)$.) Consider the homotopy

$$(1 - t)V(x) + tV_x$$

defined on $[\bar{U} - U] \times [0, 1]$. Suppose there is a point $p_0 \in \bar{U} - U$ and a number $t_0 \in (0, 1)$ such that

$$(1 - t_0)V(p_0) + t_0 v_{p_0} = 0$$

Then

$$V(p_0) = -\frac{t_0}{1 - t_0} V_{p_0} \tag{11}$$

Take the innerproduct of V_{p_0} with (11) and obtain:

$$(V(p_0), V_{p_0}) = -\frac{t_0}{1 - t_0} \|V_{p_0}\|^2 \tag{12}$$

Since $\|V_{p_0}\|^2 = 1$ and $-t_0/1 - t_0 < 0$, equation (12) contradicts (10).   □

Corollary 5.1 *If $\bar{U}$ is a ball of unit radius with center at the origin, then the index of $V(x)$ is $+1$.*

Proof. The index of the vector field V_x is $\deg[I, \bar{U}, 0]$ where I is the identity mapping.   □

Corollry 5.2 *If for each $q \in \bar{u} - U$, the vector $V(q)$ is such that*

$$(V(q), V_q) < 0$$

then the index of $V(q)$ is equal to $(-1)^n$ times the index of V_q.   □

Proof. First, $(-V(q), V_q) > 0$ and hence by the theorem, the index of $[-V(q)]$ is equal to the index of V_q. But the index of $[-V(q)]$ is $(-1)^n$ times the index of $V(q)$.

Corollary 5.3 *If the hypotheses of Corollary 6.1 or Corollary 6.2 are satisfied, then the vector field $V(p)$ has a singularity in U.*

Proof. Follows from the definition of index of a vector field and Property 1 of topological degree.   □

Brouwer Fixed Point Theorem

  Let $\bar{U}$ be the closure of a bounded open set U in R^n such that $\bar{U}$ is homeomorphic to the unit ball B^n in R^n. Suppose f is a continuous *mapping from $\bar{U}$ into $\bar{U}$. Then f has a fixed point, i.e. there is a point $p \in \bar{U}$ such that $f(p) = p$.*

Proof. Let

$$h: \bar{U} \to B_n$$

be the homeomorphism from $\bar{U}$ onto B^n given the hypothesis. We show that $g = hfh^{-1}$ has a fixed point. Suppose g does not have a fixed point on ∂B^n. Since g takes B^n into itself, then the vector field

$$G(p) = \overrightarrow{p, g(p)}$$

has index $(-1)^n$ by Corollary 6.2 and hence has a singularity, i.e. there exists $p \in B^n$ such that $p = g(p)$. $\quad\square$

By using the topological degree theory previously developed, we can give the above quick proof of the Brouwer Fixed Point Theorem. However, the Brouwer Fixed Point Theorem is independent of the degree theory in the sense that an "elementary" proof, which requires no "topological machinery," can also be given. See Alexandroff-Hopf [1945, p. 376].

Jordan Curve Theorem *Let h be a $1-1$ continuous mapping of*

$$S = \{(x, y) \mid x^2 + y^2 = 1\}$$

into R^2. Then

$$R^2 - h(S) = C_1 \cup C_2$$

where C_1, C_2 are disjoint connected sets; the set C_1 is bounded and the boundary of C_1 is $h(S)$; the set C_2 is unbounded and the boundary of C_2 is $h(S)$. If $p \in C_1$, then

$$\deg[h, B_2, p] = \pm 1 \tag{13}$$

the sign depending on the orientation of $h(S)$. If $p \in C_2$, then

$$\deg[h, B_2, p] = 0 \tag{14}$$

Proof. See Dieudonné [1960, p. 251 ff.].

The conditions (13) and (14) in the Jordan Curve Theorem as stated here are often omitted from the statement of the theorem. However we need these conditions in the proof of the Poincaré-Bendixson Theorem.

Next we define a topological degree for certain classes of mappings in infinite-dimensional spaces, i.e. Banach spaces. (Actually for the definition of this topological degree, it is sufficient to consider a linear normed space. But the spaces that occur in applications are usually Banach spaces.) It turns out that it is not possible to define a topological degree for all continuous mappings as in the finite-dimensional case. It can be shown with examples (see Cronin [1964, pp. 124-130]) that if it is assumed that a topological degree with Properties 1 and 2 has been defined for continuous mappings from a Banach space into itself, then the identity mapping (which must have topological degree ± 1) is homotopic to a constant mapping (which must have topological degree 0). This, of course, contradicts the invariance under homotopy (Property 2).

However a topological degree has been defined for a special class of mappings which often occur in analysis.

Definition. Let X be a Banach space and let f be a continuous mapping from X into itself such that if B is a bounded set in X (i.e., there exists a positive number M such that if $x \in B$, then $\|x\| < M$) then the set $F(B)$ is compact in X (i.e. each infinite subset of $F(B)$ has a limit point in X). Then F is said to be a *compact mapping* of X into itself.

It is not difficult to show that a mapping of the form $I + F$ where I is the identity and F is compact can be approximated by a sequence of mappings in finite-dimensional spaces and all mappings in this sequence which are fine enough approximations of $I + F$ have the same topological degree (i.e. Brouwer degree). This topological degree is defined to be the topological degree of $I + F$. It is usually called the Leray-Schauder degree after the two mathematicians who introduced the definition and is denoted by $\deg_{LS}[I + F, \bar{W}, p]$ where W is a bounded open set in X and $p \notin (I + F)(\bar{W} - W)$. The Leray-Schauder degree has properties exactly analogous to the Properties

1, 2 and 3 of the Brouwer degree. Most important, it turns out that many functional equations can be formulated as the study of an equation in a Banach space of X of the form

$$(I + F)x = y \qquad (15)$$

and hence the problem can be approached by investigating the Leray-Schauder degree of $I + F$. For example, it is not difficult to show that the integral equation

$$x(t) + \int_0^1 K[s, t, x(s)]ds = y(t)$$

where $y(t)$ is a continuous real-valued function on $[0, 1]$ and $K(s, t, \xi)$ is a continuous real-valued function on

$$[0, 1] \times [0, 1] \times R$$

where R is the set of real numbers, is of the form (15), i.e. the mapping

$$x(t) \to \int_0^1 K[s, t, x(s)]ds$$

is a compact mapping from $C[0, 1]$ into itself.

Many important and useful results in nonlinear analysis have been obtained by use of the Leray-Schauder degree. For a summary of some of the results up to 1963, see Cronin [1964]. Numerous applications have been obtained since then (see the Mathematical Reviews). Also the definition of Leray-Schauder degree has been extended to include larger classes of mappings. Some references to these definitions are given at the beginning of our discussion of topological degree.

Property 1. Let $\bar{W}$ be the closure of a bounded open set W in a Banach space X and F a compact mapping from $\bar{W}$ into X. Suppose $p \in X$ is such that

$$p \notin (I + F)(\bar{W} - W)$$

and suppose

$$\deg_{LS}\left[I+F,\bar{W},p\right]\neq 0$$

Then there exists $q \in W$ such that $(I+F)q = p$.

Property 2. Let $\bar{W}$ be the closure of a bounded open set W in a Banach space X and suppose that $\mathcal{F}$ is a continuous mapping from $\bar{W}\times[0,1]$ into X such that for each $\mu \in [0,1]$ the mapping

$$\mathcal{F}/(\bar{W}\times\{\mu\})$$

regarded as a mapping from $\bar{W}$ into X, is a compact mapping and such that $\mathcal{F}$ is uniformly continuous in μ, i.e. given $\varepsilon > 0$ then there exists a $\delta > 0$ such that if

$$|\mu_1 - \mu_2| < \delta$$

then for all $x \in \bar{W}$,

$$\|\mathcal{F}(x,\mu_1) - \mathcal{F}(x,\mu_2)\| < \varepsilon$$

Let $p \in X$ be such that

$$p \notin I(\bar{W}-W) + \mathcal{F}\left\{(\bar{W}-W)\times[0,1]\right\}$$

Then for each $\mu \in [0,1]$ the Leray-Schauder degree of $I + \mathcal{F}/\bar{W}\times\{\mu\}$ is defined and for all $\mu \in [0,1]$, the Leray-Schauder degree of $I + \mathcal{F}/\bar{W}\times\{\mu\}$ has the same value.

Finally we state the infinite-dimensional analog of the Brouwer Fixed Point Theorem.

Schauder Fixed Point Theorem *Let K be a bounded convex closed set in a Banach space X and F a compact mapping from K into X such that $F(K) \subset K$. Then F has a fixed point, i.e. there exists an $x \in K$ such that $F(x) = x$.*

References

Aggarwal, J. K. and Vidyasagar, M., *Nonlinear systems stability analysis*, Dowden, Hutchinson and Ross, Inc. Stroudsburg, Pennsylvania, 1977.

Ahlfors, Lars V., *Complex analysis, an introduction to the theory of analytic functions of one complex variable*, 2nd edition, McGraw-Hill, New York, 1966.

Alexandroff, P. and Hopf, H., *Topologie*, I, Springer, Berlin, 1935 (Reprinted by Edwards Brothers, Ann Arbor, Michigan, 1945).

Andronow, A. A. and Chaikin, C. E., *Theory of oscillations*, English Language Edition edited under the direction of Solomon Lefschetz, Princeton University Press, Princeton, New Jersey, 1949.

Arnold, V. I., "Proof of a theorem of A. N. Kolmogorov on the invariance of quasi-periodic motions under small perturbations of the Hamiltonian," *Russian Mathematical Surveys* 18 (1963), 9–36.

———— , *Ordinary differential equations,* translated and edited by Richard S. Silverman, M.I.T. Press, Cambridge, Massachusetts, 1973.

357

Besicovitch, A. S., *Almost periodic functions*, Dover Publications, Inc., New York, 1954.

Boyce, William and DiPrima, Richard, *Elementary differential equations and boundary value problems*, 4th edition, John Wiley & Sons, New York, 1986.

Brouwer, L. E. J., "Uber Abbildung von Mannigfaltigkeiten," *Mathematische Annalen* 71 (1912), 97–115.

Browder, Felix E. and Petryshyn, Walter, "Approximation methods and the generalized topological degree for nonlinear mappings in Banach spaces," *Journal of Functional Analysis* 3 (1969), 217–245.

Cesari, Lamberto, *Asymptotic behavior and stability problems in ordinary differential equations*, 3rd edition, Springer-Verlag, 1971.

Churchill, Ruel V. and Brown, James W., *Fourier series and boundary value problems*, 4th edition, McGraw-Hill, New York, 1987.

Coddington, Earl A. and Levinson, Norman, *Theory of ordinary differential equations*, McGraw-Hill Book Company, Inc., New York, 1955.

Courant, R. and Hilbert, D., *Methods of mathematical physics*, Volume 1, Interscience, New York, 1953.

Cronin, Jane, "Analytic functional mappings," *Annals of Mathematics* 58 (1953), 175–181.

———— "Some mappings with topological degree zero," *Proceedings of the American Mathematical Society* 7 (1956), 1139–1145.

————, "An upper bound for the number of periodic solutions of a perturbed system," *Journal of Mathematical Analysis and Applictions*, I (1960), 334–341.

———— , *Fixed Points and topological degree in nonlinear analysis,* *Mathematical Surveys,* Number 11, American Mathematical Society, Rhode Island, 1964.

———— , "A stability criterion," *Applicable Analysis* I (1971), 25–30.

———— , "Topological degree and the number of solutions of equations," *Duke Mathematical Journal* 38 (1971), 531–538.

———— , *Mathematical aspects of Hodgkin-Huxley neural theory,* *Cambridge Studies in Mathematical Biology,* Cambridge University Press, Cambridge, 1987.

Dickson, L.E., *New First Course in the Theory of Equations,* John Wiley & Sons, New York, 1939.

Dieudonne, J., *Foundations of modern analysis,* Academic Press, New York, 1960.

Field, R. J., Körös, E. and Noyes, R. M., "Oscillations in chemical systems, II. Thorough analysis of temporal oscillations in the bromate-cerium-malonic acid system," *Journal of the American Chemical Society,* 94 (1972), 8649–8664.

Field, R. J. and Noyes, R. M., "Oscillations in chemical systems, IV. Limit cycle behavior in a model of a real chemical reaction," *Journal of Chemical Physics,* 60 (1974), 1877–1884.

FitzHugh, Richard, "Mathematical models of excitation and propagation in nerve," Chapter I, *Biological engineering,* edited by Herman P. Schwan, McGraw-Hill, New York, 1969.

Friedrichs, K. O., *Lectures on advanced ordinary differential equations,* Gordon and Breach, New York, 1965.

Golubitsky, Martin and Guillemin, Victor, *Stable mappings and their singularities,* Springer-Verlag, New York, 1973.

Gomory, Ralph, "Critical points at infinity and forced oscillations," *Contributions to the Theory of Nonlinear Oscillations*, Volume III, *Annals of Mathematics Studies*, Number 36, Princeton University Press, Princeton, New Jersey, 1956.

Goodwin, B. C., *Temporal organization in cells*, Academic Press, New York, 1963.

———— , "Oscillatory behavior in enzymetric control processes," *Advances in Enzyme Regulation* (G. Weber, Ed.) 3, 1965.

Gottschalk, Walter H. and Hedlund, Gustav A., *Topological dynamics*, American Mathematical Society (Colloquium publications, Vol. 36), Providence, Rhode Island, 1955.

Hahn, Wolfgang, *Stability of motion*, Springer-Verlag, New York, 1967.

Halmos, Paul R., *Finite-dimensional vector spaces*, 2nd edition, Van Nostrand, Princeton, New Jersey, 1958.

Hartman, Philip, *Ordinary differential equations*, John Wiley & Sons, New York, 1964.

Hastings, J. W., "Are circadian rhythms conditioned?" Abstract, *Proceedings of the XI Conference of the International Society for Chronobiology*, Hanover, Germany, 1973.

Hastings, S. P. and Murray, J. D., "The existence of oscillatory solutions in the Field-Noyes model for the Belousov-Zhabotinski reaction," *SIAM Journal of Applied Mathematics* 28 (1975), 678–688.

Hirsch, Morris W. and Smale, Stephen, *Differential equations, dynamical systems, and linear algebra*, Academic Press, New York, 1974.

Hodgkin, A. L. and Huxley, A. F., "A quantitative description of membrane current and its applications to conduction and excitation in nerve," *Journal of Physiology* (1952) 117, 500–544.

Hopf, E., "Abzweigung einer periodischen Lösung von einer stationären Lösung eines differentialsystems," *Berichten der Mathematisch-Physischen Klasse der Sachsischen Akademie der Wissenschaften zu Leipzig*, XCIV, Band Sitzung vom. 19 Januar, 1942.

Hurewicz, Witold and Wallman, Henry, *Dimension theory*, revised, Princeton University Press, Princeton, New Jersey, 1948.

Ize, Jorge, "Bifurcation theory for Fredholm operators," *Memoirs of the American Mathematical Society*, Vol. 174, American Mathematical Society, Providence, Rhode Island, 1976.

Jack J. J. B., Noble, D. and Tsien, R. W., *Electric current flow in excitable cells*, Oxford University Press, 1975.

Kolmogorov, A. N. and Formin, S. V., *Elements of the theory of functions and functional analysis* (translated by Leo F. Boron), Graylock Press, Rochester, New York, 1957.

Kopell, N. and Howard, L. N., "Plane wave solutons to reaction-diffusion equations," *Studies in Applied Mathematics*, Vol. LII (1973), 291–328.

Krasnosel'ski, M. A., *Topological methods in the theory of nonlinear integral equations* (translated from Russian), Macmillan Company, New York, 1964.

LaSalle, Joseph and Lefschetz, Solonom, *Stability by Liapunov's direct method with applications*, Academic Press, New York and London, 1961.

Lefschetz, Solomon, *Differential equations: geometric theory*, 2nd edition, Interscience Publishers, New York, 1962.

Leontovich, A. M., "On the stability of the Lagrangian periodic solutions of the restricted three-body problem," *Doklady Akademii Nauk* SSSR 143 (1962), 525–528. (English translation: *Soviet Mathematics, Doklady* 3, 425–429).

Leray, Jean, "Note, Topologie des espaces abstracts de M. Banach," *Comptes Rendus Hebdomadaires des Séances de l'Académie des Sciences* (Paris) 200 (1935) 1082–1084.

Lyapunov(Liapunov), A. M., *Problème général de la stabilité du mouvement* (Reproduction of the French translation in 1907 of a Russian memoire dated 1892), *Annals of Mathematics Studies*, No. 17, Princeton University Press, Princeton, New Jersey, 1947.

Macaulay, F. S., *The algebraic theory of modular systems*, Cambridge University Press, 1916.

Malkin, I. G., *Some problems in the theory of nonlinear oscillations*, 2 volumes (AEC-tr-3766 book 1-2), U.S. Atomic Energy Commission Technical Information Service, 1959.

Marden, M., *The geometry of the zeros of a polynomial in a complex variable, Mathematical Surveys*, No. 3, American Mathematical Society, Providence, Rhode Island, 1949. Second edition, 1966.

Marsden, J. and McCracken, M., *The Hopf bifurcation and its applications, Applied Mathematical Sciences*, Volume 19, Springer-Verlag, New York, 1976.

Mawhin, J. and Rouche, N., *Équations différentielles ordinaires*, Tome 1, Tome 2, Masson et c^{ie}, Paris, 1973.

May, Robert W., *Stability and complexity in model ecosystems*, Princeton University Press, Princeton, 1973.

Maynard Smith, J., *Models in ecology*, Cambridge University Press, Cambridge, 1974.

Moser, Jürgen, *Stable and random motions in dynamical systems, Annals of Mathematics Studies* Number 77, Princeton University Press, Princeton, New Jersey, 1973.

Murray, J. D., "On a model for the temporal oscillations in the Belousov-Zhabotinsky reaction," *Journal of Chemical Physics* 61 (1974) pp. 3610–3613.

Nagumo, M., "A theory of degree of mapping based on infinitesimal analysis," *American Journal of Mathematics* 73 (1951), 485–496.

Nemytskii, V. V. and Stepanov, V. V., *Qualitative theory of differential equations*, Princeton University Press, Princeton, New Jersey, 1960.

Nirenberg, L., *Topics in nonlinear funcitonal analysis*, Courant Institute of Mathematical Sciences, New York, 1974.

Nussbaum, Roger, "The fixed point index for local condensing maps," *Annali di Matematica* 89 (1971), 217–258.

——— , "Degree theory for local condensing maps," *Journal of Mathematical Analysis and Applications* 37 (1972), 741–766.

——— , "On the uniqueness of the topological degree for k-set contractions," *Mathematische Zeitschrift* 137 (1974), 1–6.

Oatley, Keith and Goodwin, B. C., "The explanation and investigation of biological rhythms," Chapter I in *Biological rhythms and human performance*, edited by W. P. Colquhoun, Academic Press, 1971.

Poincaré, Henri, Mémoire sur les courbes définies par une équation différentielle, *Journal de Mathématiques Pures et Appliquées* (3) 7 (1881), 375–422; 8 (1882), 251–296; *Journal de Mathématiques Pures et Appliquées* (4) 1 (1885), 167–244; 2 (1886), 151–217.

——— , *Les méthodes nouvelles de la mécanique céleste*, 3 volumes, Paris, Gauthiers-Villars, 1892-99. Reprinted by Dover Publications, New York, 1958.

Royden, H. L., *Real analysis*, 2nd edition, Macmillan Company, London, 1968.

Ruelle, D. and Takens, F., "On the nature of turbulence," *Communications in Mathematical Physics* 20 (1971), 1962.

Sard, Arthur, "The measure of the critical values of differentiable maps," *Bulletin of the American Mathematical Society* 48 (1942), 883–890.

Schwartz, A. J., "A generalization of a Poincaré-Bendixson theorem to closed two-dimensional manifolds," *American Journal of Mathematics* 85 (1963), 453–458.

Scott, Alwyn C., "The electrophysics of a nerve fiber," *Reviews of Modern Physics* 47 (1975), 487–553.

Sell, George, "Periodic solutions and asymptotic stability," *Journal of Differential Equations* 2 (1966), 143–157.

Traub, Roger D. and Miles, Richard, *Neuronal networks of the hippocampus*, Cambridge University Press, Cambridge, 1991.

Tyson, John J., *The Belousov-Zhatotinski reaction, Lecture Notes in Biomathematics*, Springer-Verlag, Berlin, New York, 1976.

Van der Waerdon, B., *Moderne algebra*, Volume II, 2nd edition, Springer, Berlin, 1940. Translation into English by Fred Blum, F. Ungar Publishing Co., New York, 1949.

Winfree, Arthur T., "Biological rhythms and the behavior of coupled oscillators," *Journal of Theoretical Biology* 16 (1967), 15–42.

Wintner, Aurel, *The analytical foundations of celestial mechanics*, Princeton University Press, Princeton, New Jersey, 1947.

Index

– W –